Oxford Chemistry Primers

SERIES EDITOR
STEPHEN G. DAVIES
The Dyson Perrins Laboratory, University of Oxford

OXFORD CHEMISTRY PRIMERS

Physical Chemistry Editor
RICHARD G. COMPTON
Physical Chemistry Laboratory
University of Oxford

Founding Editor and
Organic Chemistry Editor
STEPHEN G. DAVIES
The Dyson Perrins Laboratory
University of Oxford

Inorganic Chemistry Editor
JOHN EVANS
Department of Chemistry
University of Southampton

Reactive Intermediates

Christopher J. Moody

Department of Chemistry, Loughborough University of Technology
and

Gordon H. Whitham

The Dyson Perrins Laboratory and Pembroke College, University of Oxford

OXFORD   NEW   YORK   TOKYO
OXFORD UNIVERSITY PRESS

Oxford University Press, Walton Street, Oxford OX2 6DP

Oxford New York
Athens Auckland Bangkok Bombay
Calcutta Cape Town Dar es Salaam Delhi
Florence Hong Kong Istanbul Karachi
Kuala Lumpur Madras Madrid Melbourne
Mexico City Nairobi Paris Singapore
Taipei Tokyo Toronto

and associated companies in
Berlin Ibadan

Oxford is a trade mark of Oxford University Press

Published in the United States
by Oxford University Press Inc., New York

© C. J. Moody and G. H. Whitham, 1992
Reprinted 1995 (with corrections)

A catalogue record for this book is available from the British Library

Library of Congress Cataloging in Publication Data
(data available)
ISBN 0 19 855672 1 (Pbk)

Printed in Great Britain by Information Press, Eynsham, Oxford

Series Editor's Foreword

Neutral reactive intermediates have progressed from being mechanistic curiosities to being established and reliable tools for organic synthesis, and hence form an essential part of undergraduate chemistry courses.

Oxford Chemistry Primers have been designed to provide concise introductions relevant to all students of chemistry and contain only the essential material that would usually be covered in an 8–10 lecture course. In this eighth primer of the series Chris Moody and Gordon Whitham discuss from first principles the preparation and properties of neutral reactive intermediates with particular emphasis on their use in synthesis. This primer will be of interest to apprentice and master chemist alike.

Stephen G. Davies
The Dyson Perrins Laboratory, University of Oxford

Preface

The use of neutral reactive intermediates (radicals, carbenes, nitrenes, and arynes) in organic synthesis has increased dramatically over the past 10–15 years. It is now commonplace for organic chemists to use reactions which proceed via radical intermediates; these processes can be as equally efficient and high yielding as those involving polar intermediates such as carbanions. Carbenes, too, particularly in the form of transition-metal carbenoids, as well as nitrenes and arynes, are increasingly used in synthesis.

Although reactions involving neutral reactive intermediates do appear in standard undergraduate texts, the discussion is usually very brief, and does not emphasize the synthetic potential of such species. This short book aims to fill that gap by providing an up-to-date concise account of the status of reactive intermediates in organic synthesis. The intermediates are discussed in turn, and although the emphasis is on preparatively useful reactions, particularly in the chapter on radicals, the chapters on carbenes, nitrenes, and arynes do contain short accounts of the fundamental properties of the intermediates.

Finally, we thank Dr Russ Bowman for his helpful comments on Chapters 3–5.

Loughborough
Oxford
January 1992

C. J. M.
G. H. W.

Contents

1 Introduction

Many of the reactions of organic chemistry proceed by way of reactive intermediates according to the following schematic equation.

$$\text{starting material} \underset{k_{-1}}{\overset{k_1}{\rightleftharpoons}} \text{intermediate(s)} \underset{k_{-2}}{\overset{k_2}{\rightleftharpoons}} \text{product(s)}$$

In most of the cases of interest to us in this book, $k_2 > k_1$ otherwise the intermediate would represent an isolable compound or a species in rapid equilibrium with reactants. In general, reactive intermediates correspond to a relatively shallow dip in a free energy versus reaction co-ordinate diagram and they can either proceed to products faster than returning to starting material, i.e. $k_2 > k_{-1}$ or vice-versa, $k_{-1} > k_2$. Much effort has been expended in certain famous test cases such as the 'non-classical' carbonium ions in deciding whether an intermediate actually exists or not. It is usually considered that a reactive intermediate is significant if the depth of the free energy well containing it is sufficient to prevent every molecular vibration along the reaction co-ordinate proceeding back to reactants or forward to products.

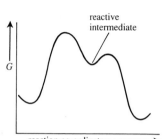

Reaction profile showing reactive intermediate where $k_2 > k_{-1}$

benzyne

Many of the reactive intermediates of organic chemistry are charged species, such as carbocations (carbenium ions) or carbanions but there is an important sub-group of formally neutral electron-deficient reactive intermediates. In examples containing carbon as the reactive centre these can be either trivalent, with a single non-bonding electron, i.e. a carbon centered radical, or divalent with two non-bonding electrons, i.e. a carbene. A comparison of various neutral reactive intermediates and their relationship to corresponding cations and anions is shown in the table on page 2. This book is concerned with the chemistry of radicals, carbenes and nitrenes (the nitrogen analogues of carbenes), and also with another type of neutral reactive intermediate, benzynes. The latter, highly strained cyclic benzenoid acetylenes, have many chemical affinities to carbenes.

This is not the place to describe in detail the experimental basis on which the involvement of reactive intermediates in specific reactions has been established but it is appropriate briefly to mention the sort of evidence that has been found useful in this respect.

Probably one of the most direct ways in which reactive intermediates can be inferred in a particular reaction is by a kinetic study. Trapping the intermediate with an appropriate reagent can also be very valuable, particularly if it can be shown that the same products are produced in the same ratios when the same postulated intermediate is formed from different precursors.

A classic example of the combined use of kinetic and product trapping studies is that of Hine and co-workers, described on p.34, on the hydrolysis of chloroform under basic conditions. The observation that chloroform undergoes deuterium for hydrogen exchange (in D_2O) faster than hydrolysis,

Relationship between reactive intermediates

	C	N	O
-onium ion	R_5C^+ *carbonium ion*	R_4N^+ *ammonium ion*	R_3O^+ *oxonium ion*
neutral molecule	R_4C	R_3N	R_2O
anion	R_3C^- *carbanion*	R_2N^- *amide anion*	RO^- *alkoxide*
radical	$R_3C\cdot$ *carbon radical*	$R_2N\cdot$ *aminyl radical*	$RO\cdot$ *oxyl radical*
-enium ion	R_3C^+ *carbenium ion*	R_2N^+ *nitrenium ion*	RO^+ *oxenium ion*
-ene	$R_2C:$ *carbene*	$RN:$ *nitrene*	$:O:$ *oxene*

and further that the rate of hydrolysis is retarded by addition of chloride ion is strong evidence in favour of the mechanism shown.

$$HO^- + CHCl_3 \underset{\text{fast}}{\rightleftharpoons} H_2O + {}^-CCl_3 \underset{\text{slow}}{\rightleftharpoons} :CCl_2 + Cl^-$$

$$\downarrow$$

$$\text{products}$$

Further circumstantial evidence for dichlorocarbene formation is provided by trapping experiments, e.g. with alkenes, giving 1,1-dichlorocyclopropanes as products (p.35).

Spectroscopic methods have also provided valuable evidence for the intermediacy of transient species, particularly where combined with flash photolysis when high concentrations of the intermediate can be built up for UV detection, or by using matrix isolation techniques when species such as *o*-benzyne can be detected and their IR spectra obtained (p.69).

In the case of transient species with unpaired electrons such as free radicals, and the triplet states of carbenes or nitrenes, electron paramagnetic resonance ('ESR') spectroscopy can provide unique evidence about the structure of the intermediate. Useful information about intermediates in reactions involving radical pair coupling can also be obtained by a technique known as chemically induced dynamic nuclear polarisation (CIDNP). However detailed discussions of ESR and CIDNP are outside the scope of this book and for further information suitable text books on physical organic chemistry or the references for further reading should be consulted.

Further reading

Kinetic Evidence for Reactive Intermediates, R.Huisgen, *Angew. Chem. Int. Ed.Engl.,* 1970, **9**, 751.

Structure of Free Radicals by ESR Spectroscopy, H. Fischer, in *Free Radicals*, ed. J. K. Kochi, J. Wiley, New York, 1973, Vol. 2, 435.

Chemically Induced Dynamic Nuclear Polarisation, H. R. Ward, in *Free Radicals,* ed. J. K. Kochi, J. Wiley, New York, 1973, Vol. 1, 239.

2 Radical reactions in synthesis

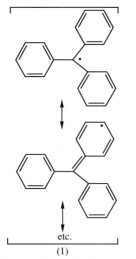

Triphenylmethyl radical, stabilised by delocalisation

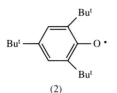

(2)

Persistent radical with sterically shielded radical centre

For many years the chemistry of free radicals was very much the province of mechanistic and physical organic chemists and applications to synthesis did not extend much beyond the occasional use of Kolbe electrolysis or oxidative coupling of phenols. More recently the situation has changed, and the realisation that radical methods are often compatible with a range of functional groups, without further protection, has led to an increased interest in the use of radicals in synthesis. The appearance of a book by B.Giese devoted to the subject and some valuable review articles, (cited at the end of this chapter), have also led to an increased awareness of the potential of radical based synthetic procedures.

In this chapter we shall use the term 'radical' rather than the older term 'free radical', meaning an atom or molecule which possesses one or more unpaired electrons.

The fundamental principles of homolytic reactions and radical chemistry are dealt with in basic textbooks and they will not be repeated in detail here. Since our prime consideration is to emphasise synthetic aspects, we shall not be concerned particularly with stable or persistent radicals, such as triphenylmethyl (1) and the substituted phenoxy radical (2). For similar reasons, we shall devote relatively little attention to the chemistry of very reactive radicals, such as chlorine atoms, since their highly exothermic reactions lead to poor selectivity.

The overall intention of this chapter is to concentrate on those reactions where preparatively useful yields can be obtained using radical processes in solution.

2.1 Methods of initiation of radical reactions

Since, as we shall see, most of the important radical reactions proceed by a chain mechanism, the crucial initiation step requires generation of radicals from an appropriate precursor. This section deals with the main types of reaction leading to the production of radicals, i.e. thermal cleavage of compounds with weak bonds, photochemical cleavage of compounds with weak bonds, and electron transfer processes.

Thermal cleavage of compounds with weak bonds

There are a number of compounds that contain relatively weak bonds, i.e. with bond dissociation energies (BDE) less than about 160kJmol^{-1}, that undergo homolysis at a convenient rate at temperatures below 150°C.

An important group of such compounds is the diacylperoxides (3),where the already weak bond of a simple peroxide (for example the O–O bond of MeO–OMe has a BDE of 155 kJmol^{-1}) is further weakened by resonance stabilisation of the acyloxy radical (4) produced on homolysis (equation 2.1). Thus the BDE of the O–O bond in diacetyl peroxide (3,R = Me) is 126 kJmol^{-1}.

$$\text{(3)} \quad \xrightarrow{\text{heat}} \quad 2 \text{ (4)} \quad \xrightarrow{-CO_2} \quad R\cdot \text{ (5)} \qquad 2.1$$

Decarboxylation of the acyloxy radical (4) may or may not occur readily depending on the nature of R. For R = alkyl the rate of loss of carbon dioxide is relatively fast and the only important products are those derived from the alkyl radicals thereby produced. With R = aryl, decarboxylation is slower and products derived either from $ArCO_2\cdot$ or from $Ar\cdot$ or from both can be obtained depending on the rates of the subsequent product determining reactions.

In practice, diacyl peroxides (3) with R = aryl, e.g. dibenzoyl peroxide (3, R = Ph), are often used as initiators for radical reactions since they dissociate at a convenient rate (half life about 1 h at 90°C) while the alkyl analogues tend to be avoided owing to their explosive nature.

Of the simpler dialkylperoxides, di-t-butylperoxide (6) (O–O BDE ~ 155 kJmol^{-1}; half life ~1h at 150°C) is occasionally used as an initiator. Again a complicating feature arises: the derived t-butoxy radicals (7) can either react directly or undergo fragmentation to methyl radicals and acetone depending on the rates of the subsequent reactions (equation 2.2).

$$\text{(6)} \quad \xrightarrow{\text{heat}} \quad 2 \text{ (7)} \quad \longrightarrow \quad + \; CH_3\cdot \qquad 2.2$$

Another important group of compounds often used as precursors of radicals are the azo-compounds (8); these dissociate on heating, with loss of nitrogen, to give radicals at temperatures which depend upon the nature of R. In general the more stable the radical R· the more readily dissociation occurs (equation 2.3).

$$R\text{-}N{=}N\text{-}R \quad \xrightarrow{\text{heat}} \quad N_2 \; + \; 2\,R\cdot \qquad 2.3$$
$$\text{(8)}$$

Azomethane (8, R = Me) decomposes to nitrogen and methyl radicals in a reaction with an activation energy (E_a) of about 210 kJmol^{-1}, while (8, R = Ph$_2$CH), which leads to resonance stabilised diphenylmethyl radicals, has an activation energy for decomposition of 110kJmol^{-1}. One of the most commonly used radical precursors is azobisisobutyronitrile (AIBN) (9). The ease of decomposition of AIBN (half life ~ 1·0 h at 80°C) reflects the fact that the resulting radicals are both tertiary and resonance stabilised as shown.

Delocalisation of the unpaired electron in the radicals from AIBN

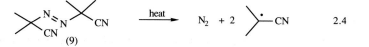

$$\text{(9)} \xrightarrow{\text{heat}} N_2 + 2 \overset{\cdot}{\diagup}\!\!-CN \qquad 2.4$$

Other examples of the thermal dissociation of appropriately constituted radical precursors will be encountered later. They all have in common the homolytic cleavage to give stabilised radicals, often accompanied by the formation of small spin-paired molecules such as nitrogen or carbon dioxide, which helps to favour the overall decomposition on entropy grounds as in the case of the azoalkanes (8).

Photochemical cleavage of compounds with weak bonds

If it is desirable to effect the dissociation of an initiator into radicals at room temperature or below, e.g. because the reaction products are thermally unstable, it is often practicable to use photolysis rather than thermolysis provided the compound used absorbs light of an appropriate wavelength. Since a quantum of light of wavelength 300 nm corresponds to about 400 kJmol^{-1} it is energetically feasible to cleave weak covalent bonds by irradiation with visible or UV light.

Well known examples include the halogens, e.g. bromine (BDE = 193 kJmol^{-1}) is a source of bromine atoms (equation 2.5). The azoalkanes and dialkyl- and diarylperoxides discussed above as thermal radical precursors, also absorb light above 300nm and give good radical yields on irradiation.

$$Br_2 \xrightarrow{h\nu} [Br_2]^* \longrightarrow 2\,Br\cdot \qquad 2.5$$

There is another type of photoinitiated radical process which occurs when the electronically excited state of a molecule undergoes an inter- or intra-molecular atom transfer. By way of an example, the excited (triplet) state of benzophenone (10), which can be considered as a sort of alkoxy radical since one of the lone pair electrons on oxygen has been promoted to the antibonding π^*–orbital, can react with the secondary alcohol benzhydrol (11) by abstraction of the hydrogen atom α to OH. The resulting so-called ketyl radicals (12), being stabilised and relatively long lived, can build up to a sufficiently high concentration to allow the product of radical combination, benzpinacol (13), to be formed in good yield (equation 2.6). Ketyl radicals such as (12) are important intermediates in the photochemistry of carbonyl compounds but reactions of the type shown will not be emphasised in this chapter since they are not relevant to the main theme.

Electron transfer processes

Reactions of this type involve one electron oxidation or reduction leading to the generation of radicals from spin-paired precursors. Examples include (i) the electrochemical (anodic) oxidation of carboxylate anions (14) leading to acyloxy radicals, which as indicated in equation 2.1 can undergo decarboxylation to give alkyl radicals, the basis of the Kolbe reaction (equation 2.7), and (ii) the reduction of hydrogen peroxide by Fe(II) to give hydroxy radicals (equation 2.8), and the analogous reduction of alkyl hydroperoxides to alkoxy radicals.

$$RCO_2^- \xrightarrow{-e} RCO_2 \cdot \xrightarrow{-CO_2} R\cdot \longrightarrow R-R \qquad 2.7$$
(14)

$$HO-OH \xrightarrow{Fe^{II}} OH^- + HO\cdot + Fe^{III} \qquad 2.8$$

2.2 Main types of radical reaction

It is convenient at this stage to list and briefly discuss the main types of elementary reactions of radicals that commonly occur. Some indication of the rates at which such reactions proceed in solution will be given since it can help in synthetic planning to have an idea of the relative rates of processes that may be in competition.

Radical combination

The combination of two radicals to give a covalent molecule is an important termination step for chain reactions since it destroys radicals. It does not often play a significant preparative role. The rate of reaction is proportional to the product of the concentration of two species if the radicals are different (or the square of the concentration of one if they are the same). Since the radicals are usually transient and present in very low concentration the overall rate of reaction can be low. However, for relatively stable, long lived, radicals, and for cases where relatively high local concentrations of radicals occur, e.g. at the anode in the anodic oxidation of carboxylate ions (the Kolbe reaction, equation 2.7) or for radical pairs generated within a solvent cage, it can become a viable pathway. Rate *constants* are often high, reflecting the low activation energy for spin-pairing, and with reactive radicals these rate constants are close to the expected value for diffusion controlled reactions (*c.* $10^9 M^{-1}s^{-1}$). For example a value of $5 \times 10^7 M^{-1}s^{-1}$ is found for combination of two $Cl_3C\cdot$ radicals at 24°C.

Radical abstraction (displacement)

This is a common and important step in many chain reactions and often involves attack on a hydrogen atom, e.g. the hydrogen abstraction step in the chlorination of methane (equation 2.9), or on a halogen atom, e.g. the reaction of tributyltin radicals with an alkyl halide (equation 2.10).

$$Cl \cdot \ + \ H-CH_3 \longrightarrow Cl-H \ + \ CH_3 \cdot \qquad\qquad 2.9$$

$$Bu_3Sn \cdot \ + \ BrR \longrightarrow Bu_3SnBr \ + \ R \cdot \qquad\qquad 2.10$$

Radical reactions of this type can be considered as displacement reactions, in which the radical effects back side attack on the atom undergoing displacement, with formation of a new radical. Sometimes the symbol S_H2 is applied to such reactions, meaning 'substitution, homolytic, bimolecular' by analogy with the time honoured terminology for nucleophilic substitution. In general a radical displacement will be favourable if the bond being formed is stronger than the bond broken, and rate constants can be quite high, e.g. $10^4 M^{-1}s^{-1}$, for hydrogen abstraction from the methyl group of toluene by *t*-butoxy radicals (equation 2.11). In the latter example, attack at the weakest C–H bond leading to the resonance stabilised benzyl radical rather than at one of the aromatic C–H bonds should be noted.

Atom abstraction reactions may occur intramolecularly if a sterically favourable transition state can be attained. This turns out to be true for 1,5-hydrogen atom transfers where a six-membered transition state is involved, as shown schematically in equation 2.12 for the rearrangement of an alkoxy radical of appropriate chain length.

One electron arrows or 'fish-hooks' as used in equation 2.12 are a convenient way of showing the movement of unpaired electrons akin to the use of normal arrows to show electron pair movements

$$Bu^tO \cdot \ + \ CH_3Ph \longrightarrow Bu^tOH \ + \ \cdot CH_2Ph \qquad\qquad 2.11$$

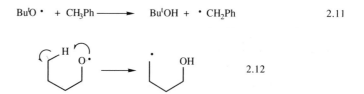

$$2.12$$

Radical displacement reactions in which attack occurs at a carbon atom rather than at a halogen or hydrogen atom are not generally observed. This is presumably because the transition state would suffer far more steric strain than, for example, the transition state for a competing hydrogen abstraction.

Radical addition to multiple bonds

The addition of a radical to a double or a triple bond is a common elementary step in radical-chain addition to multiple bonds, such as the long known peroxide catalysed 'anti Markownikow' addition of HBr to alkenes. Here the key step is the addition of a bromine atom to the double bond and the orientation found in addition to an unsymmetrically substituted alkene is governed by attack at the more sterically available site. Although early explanations of this regiochemical control tended to emphasise the relative stability of the two possible radicals formed; current views tend to favour an early transition state with steric factors being the main reason for attack at the less substituted end of the double bond. In addition, as we shall see later, there may also be a favourable polar effect influencing this mode of addition with electrophilic radicals such as bromine atoms (Section 2.4)

Rate constants for addition of radicals to double bonds are often of the order of 10^5–$10^6 M^{-1}s^{-1}$, reflecting a preparatively useful process, although they vary over quite a range depending on the nature of the attacking radical and the alkene. Some additions are reversible, e.g. addition of thiyl radicals to alkenes (equation 2.13), and the reaction has been used preparatively for isomerisation of 1,2-disubstituted (Z)-alkenes to a thermodynamic equilibrium mixture containing a predominance of the (E)-isomer.

ArS • + [alkene]R ⇌ ArS[alkene]R   2.13

Addition is also the key step in homolytic aromatic substitution, but in general this process is not particularly useful synthetically, since substituted aromatics tend to give mixtures of isomers. However in some intramolecular cases, where the site of attack is pre-determined by the ring size of the particular intermediate, the reaction has some synthetic potential.

Radical polymerisation is an important process initiated by radical addition to alkenes. It will only feature incidentally in this chapter since the emphasis is on the synthesis of relatively small molecules.

Fragmentation (β-elimination)

Fragmentation is the microscopic reverse of radical addition, as just mentioned. Some examples have already been encountered in the decarboxylation of acyloxy radicals (equation 2.1) and the β-elimination of t-alkoxy radicals to carbonyl compounds and alkyl radicals (equation 2.2). On the whole such fragmentations will be favoured when the multiple bond formed has a high bond energy (e.g. C=O) in the case of t-alkoxy radicals, or where the starting radical is strained, e.g. the ring opening of cyclopropyl methyl radicals (equation 2.14). The rate constant for the latter reaction is high, $1.0 \times 10^8 \text{s}^{-1}$ at 25°C, and the reaction has been proposed as a 'radical clock', i.e. a standard fast reaction of known rate constant against which rates of other competing reactions of the precursor radical or product radical can be measured.

Decarboxylation of acyloxy radicals

[structure]• → [structure]=   2.14

Rearrangements

Intramolecular versions of atom abstraction reactions, already mentioned, intramolecular addition reactions leading to cyclisation and intramolecular fragmentations, again briefly mentioned above, constitute pathways for rearrangement of suitably constituted radicals. Several examples of each of these processes will be encountered later in section 2.5.

There are also certain rearrangements involving 1,2-shifts of the generalised type leading from (14) to (15) (equation 2.15). In contrast to carbonium ion chemistry, where 1,2-alkyl (and hydride) shifts are common, 1,2-alkyl shifts in radicals (equation 2.15, X = alkyl) are rare. The few apparent cases are likely to be examples of β-elimination followed by re-addition the other way round. However, in cases where X is a multiply bonded group such as an aryl, vinyl, or carbonyl group, rearrangements corresponding to equation 2.15

are well known. An intramolecular addition–elimination pathway is likely to be operative, as shown for the case of phenyl migration in equation 2.16. Similarly, groups such as halogen with relatively low lying empty atomic orbitals have been found to undergo a 1,2-shift; again some sort of bridged intermediate is likely (equation 2.17)

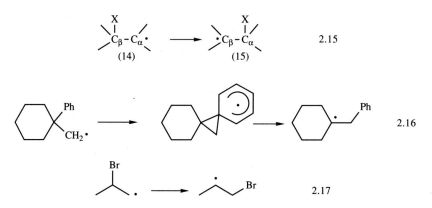

2.15

2.16

2.17

Disproportionation

Radical combination has already been mentioned. The other main radical–radical reaction is disproportionation in which atom transfer between two radicals occurs leading to two spin paired molecules. Again, as for combination, this is one of the ways of terminating a chain process and it is only preparatively important when high local concentrations of radicals can be generated. An example is the disproportionation of aliphatic ketyl radicals (16) obtained in relatively high concentration by flash photolysis, e.g. from acetone and isopropanol (equation 2.18). This is a neat way of generating non-stabilised enols in solution.

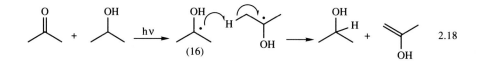

2.18

2.3   Chain reactions

For reasons already mentioned, only a small number of the free-radical reactions used in synthesis involve radical coupling as the product forming step and the majority of synthetically useful radical reactions in solution are chain processes. It is thus instructive to summarise the main features which characterise successful chain reactions.

A typical example is the much exploited reduction of alkyl bromides (or iodides) by tributyltin hydride (tributylstannane) which can be induced, for example, on heating with the initiator AIBN (9) in benzene solution. The mechanism shown in equations 2.19–2.25 is proposed. The three main reaction types are typical of radical chain reactions in general: the initiation stage, leading to formation of the reactive radicals; the propagation stage, accounting for the main overall reaction; and the termination stage, in which radicals are consumed to give spin paired molecules. In the case under discussion initation consists of thermal dissociation of AIBN to isobutyronitrile radicals which then abstract hydrogen from the weak Sn–H bond to give the key $Bu_3Sn\cdot$ radicals (equations 2.19 and 2.20). Reaction of $Bu_3Sn\cdot$ radicals with the alkyl bromide to give alkyl radicals is favoured by the relatively high Sn–Br bond energy (equation 2.21), the second propagation step (equation 2.22) is favoured by the relatively weak Sn–H bond broken, coupled with the relatively strong C–H bond formed. Note that other possible atom abstraction reactions involving $Bu_3Sn\cdot + Bu_3SnH$ and $R\cdot + RBr$ are irrelevant since they lead merely to regeneration of starting radicals. The crucial product forming steps are thus reactions 2.21 and 2.22 and the cyclic nature of the propagation sequence is emphasised by the representation in equation 2.26. The three possible combination reactions leading to comsumption of $Bu_3Sn\cdot$ and $R\cdot$ radicals constitute the possible termination reactions (reactions 2.23 to 2.25).

$$AIBN \xrightarrow{\Delta} Me_2\overset{\bullet}{C}CN \qquad\qquad 2.19$$

$$Me_2\overset{\bullet}{C}CN + Bu_3SnH \longrightarrow Me_2CHCN + Bu_3Sn\cdot \qquad 2.20$$
initiation

$$Bu_3Sn\cdot + RBr \longrightarrow Bu_3SnBr + R\cdot \qquad 2.21$$

$$R\cdot + Bu_3SnH \longrightarrow RH + Bu_3Sn\cdot \qquad 2.22$$
propagation

$$2\,Bu_3Sn\cdot \longrightarrow Bu_3SnSnBu_3 \qquad\qquad 2.23$$

$$2\,R\cdot \longrightarrow R\text{-}R \qquad\qquad 2.24$$
termination

$$Bu_3Sn\cdot + R\cdot \longrightarrow Bu_3SnR \qquad\qquad 2.25$$

$$2.26$$

Some of the salient kinetic features of viable radical chain reactions are (i) rate constants for primary initiation steps, e.g. equation 2.19, are relatively low (e.g. for equation 2.19 $k_i\sim 10^{-5}s^{-1}$); (ii) rate constants for propagation steps are relatively high ($k > 10^2\,M^{-1}s^{-1}$); (iii) rate constants for radical–radical termination steps are very high, approaching diffusion control ($k \sim 10^9 M^{-1}s^{-1}$); (iv) the overall rate of reaction is proportional to the square root of the rate of the primary initiation step; and (v) the overall activation energy is dominated by the activation energy of this latter step. The consequences are that typical radical concentrations during reaction are of the order of $10^{-7}M$, thus the actual rates of the bimolecular termination steps are low despite the high rate constants. Also, chain lengths, i.e. the ratio of the number of propagation steps to termination steps, are often of the order of tens of thousands. Clearly there are important limitations to the types of radical reactions that can participate in chain sequences and these requirements are more stringent where a more complex series of propagation steps, with more than two stages in the cycle, is involved. As we shall see later, an important case relevant to the above example (equation 2.26) is that where the first formed radical R· can rearrange to a new radical R'·. The concentration of Bu₃SnH then becomes critical to the overall outcome of the reaction.

2.4 Polar effects in radical reactions

One other general feature of radical reactions that cannot be too strongly emphasised is their marked susceptibility to polar effects in both reactants and reagents. This aspect of reactions between nominally uncharged species, which is at first sight surprising, has far reaching consequences.

Clear indications of the operation of polar effects can be found in certain hydrogen abstraction reactions. For example, propanoic acid (17) could undergo reaction by hydrogen abstraction from either the α or the β position (the O–H bond is not involved owing to its significantly higher bond energy compared to C–H).

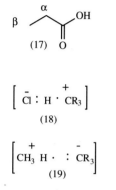

(17)

In the reaction of propanoic acid with chlorine atoms (Cl·) the relative rate constants for attack at $C_\alpha H$ versus $C_\beta H$, per hydrogen atom, are in the ratio $k_\alpha / k_\beta = 0.03$. The carboxyl group is electron withdrawing and the electronegative chlorine atom (electrophilic radical) preferentially attacks the more electron-rich C–H bond of the methyl group even though the α radical might be expected to be the more stable owing to resonance stabilisation. In contrast, for the reaction of propanoic acid with the more electron rich methyl radicals (Me·), selective attack at the more electron deficient bond α to the carboxyl group occurs ($k_\alpha / k_\beta = 7.8$). Apparently, in the transition state for chlorine atom attack there is significant electron transfer towards the electronegative chlorine leaving the carbon atom relatively positively charged, which might be described in resonance terms as a contribution from canonical (18) to the transition state. In the case of methyl radical attack there is some electron transfer towards the carbon of the C–H bond being attacked, which could be expressed as a contribution by canonical (19) to the transition state.

Typical electrophilic radicals include Hal·, RO·, RS·, RSO₂·, Cl₃C·, R·CO·CH₂·, RO₂C·CH₂·, and RSO₂CH₂·;typical nucleophilic radicals are R₃C·, ArCH₂·, R₂(R'O)C·, and RCO· .

A further illustration of polar effects is provided by the effect of substituents on the rate of addition of radicals to substituted alkenes. Relative rate constants (k_{rel}) for addition of cyclohexyl radicals, which are nucleophilic, (equation 2.27) vary with the substituents (X) as follows: CHO, 34; CO₂Me, 6·7; Ph, 1·0; OAc, 0·016; *n*-C₄H₉, 0·004. Clearly the more powerfully electron withdrawing X is, the faster is the addition, with rates spanning a range of nearly 10^4. Conversely it is found that electrophilic radicals add preferentially to electron-rich olefins substituted with electron donating substituents.

$$\begin{bmatrix} \bar{Cl} : H \cdot \overset{+}{C}R_3 \end{bmatrix}$$

(18)

$$\begin{bmatrix} \overset{+}{CH_3} \ H \cdot \ : \bar{C}R_3 \end{bmatrix}$$

(19)

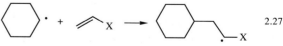

2.27

Polar effects such as these have important consequences in polymer chemistry leading to a strong tendency for formation of an alternating polymer chain when, for example, styrene and acrylonitrile are copolymerised. In the growing polymer the nucleophilic benzylic radical (20) tends to add preferentially to acrylonitrile, as the electron deficient monomer, leading to the electrophilic radical (21). The latter tends to undergo addition to styrene as the more electron rich monomer giving radical (22), and so on (equation 2.28).

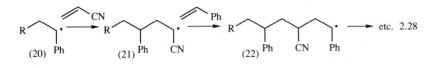

2.5 Examples of radical reactions in synthesis

We shall now examine a number of examples of radical reactions in a synthetic context, trying to illustrate the general points already made. Inevitably the choice will be highly selective and more specialist books and reviews should be consulted for a comprehensive treatment.

A particular feature of radical reactions in synthesis is their inherent compatibility with certain functional groups. In carbanion chemistry it is nearly always necessary to protect hydroxy and other acidic groups to prevent formation of alkoxides, etc. In free radical chemistry O–H cleavage is not usually a problem, owing to the high O–H bond energy, and protection is rarely necessary.

As already indicated, the main problem in the development of radical based syntheses is the selection of appropriate reactions so that the elementary steps proceed at sufficiently fast rates to sustain chain reactions. Sometimes the relevant rate constants are known or can be estimated by analogy, but successful procedures are often the result of considerable experimentation in varying reagent concentrations and reaction conditions.

Radical anion and radical cation reactions will not be emphasised in this survey since they are more appropriately dealt with in the context of reduction and oxidation.

Intermolecular radical additions to multiple bonds

A typical example is the tributyltin hydride promoted addition of alkyl radicals, generated from alkyl halides, to alkenes (equation 2.29, where W represents an electron withdrawing group).

$$RHal + \underset{W}{\diagup\!\!\!\!\diagdown} \xrightarrow[\text{AIBN}/\Delta]{Bu_3SnH} R\diagdown\!\!\!\!\diagup_W + Bu_3SnHal \qquad 2.29$$

This reaction is fairly general for primary, secondary, and tertiary alkyl bromides and iodides. One illustration is shown in equation 2.30.

$$\text{Bu}^t\text{Br} + \quad \diagup\diagdown_{\text{CN}} \quad \xrightarrow[\text{AIBN}/\Delta]{\text{Bu}_3\text{SnH}} \quad \text{Bu}^t \diagdown\diagup\diagdown^{\text{CN}} + \text{Bu}_3\text{SnBr} \qquad 2.30$$

In comparison with the tributyltin hydride reduction of alkyl halides discussed earlier (Section 2.3), there is now an extra link in the chain mechanism (equation 2.31).

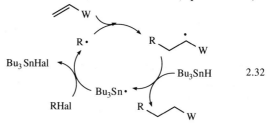

$$\text{R} \cdot + \underset{(23)}{\diagup\diagdown_{\text{W}}} \longrightarrow \underset{(24)}{\text{R}\diagdown\diagup^{\cdot}_{\text{W}}} \xrightarrow{\text{Bu}_3\text{SnH}} \text{R}\diagdown\diagup\diagdown_{\text{W}} + \text{Bu}_3\text{Sn} \cdot \quad 2.31$$

The overall propagation sequence has now been expanded from that shown in equation 2.26 to the one summarised below (equation 2.32).

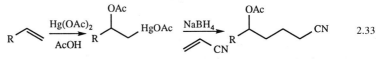

2.32

Clearly a number of conditions must be fulfilled for the process to be efficient. The radical R· must add to the activated alkene (23) faster than it reacts (by hydrogen transfer resulting in overall reduction) with Bu_3SnH, but the adduct radical (24) must react with Bu_3SnH faster than (24) reacts with another molecule of activated olefin (23) leading to polymerisation. These conditions are favoured when R· is a nucleophilic radical and the alkene is electron deficient (with W a powerfully electron withdrawing group) thereby making the adduct radical (24) an electrophilic radical. The reaction is also favoured if a high concentration of the activated alkene (23) is maintained, e.g. by working in excess, and the Bu_3SnH concentration is kept low, e.g. by slow addition during the reaction.

An alternative way of generating alkyl radicals for intermolecular addition reactions is by sodium borohydride reduction of alkyl mercurials. Many compounds of the type RHgX give alkyl radicals via radical chain decomposition of RHgH produced *in situ* by borohydride reduction and these radicals can be trapped by activated alkenes. An example is shown in equation 2.33 where the mercurial used was obtained by acetoxymercuration of a terminal alkene.

$$\text{R}\diagup\diagdown \xrightarrow[\text{AcOH}]{\text{Hg(OAc)}_2} \text{R}\overset{\text{OAc}}{\underset{}{\diagdown}}\diagup\text{HgOAc} \xrightarrow[\underset{\text{CN}}{\diagup\diagdown}]{\text{NaBH}_4} \text{R}\overset{\text{OAc}}{\underset{}{\diagdown}}\diagup\diagdown\diagup_{\text{CN}} \qquad 2.33$$

In some ways the mercurial approach is complementary to the $\text{RHal}/\text{Bu}_3\text{SnH}$ procedure but where direct comparisons are available, e.g. with alkyl radicals derived from $\text{RHgHal}/\text{NaBH}_4$, the Bu_3SnH route is preferable, in terms of yields and availability of starting materials.

An important type of radical precursor, which has recently been developed by Barton and shown to have wide applicability, is a thiohydroxamic ester (26). These compounds, readily prepared from the sodium salt of the thiohydroxamic acid (25) and the appropriate acid chloride, undergo radical chain decarboxylative rearrangement on heating or irradiation via the propagation sequence shown in equation 2.34.

Factors favouring the chain mechanism here include: addition to the weak C=S, formation of the aromatic pyridine ring, and fragmentation with loss of CO_2

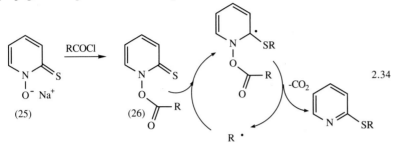

2.34

Alkyl radicals (R·) formed by photolysis of (26) can be trapped by electron deficient alkenes leading, for example, to formation of the adduct (28) starting from the thiohydroxamic ester (27) (equation 2.35).

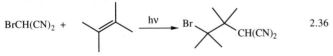

2.35

Adducts of this type (28) can be further transformed via the manipulations of organosulphur chemistry leading, for example, to α,β-unsaturated esters via oxidation to sulphoxide followed by thermal elimination, or to saturated esters by reduction.

Addition of electron deficient radicals to electron rich alkenes is also possible and some of the above methods have been used to this end. Many of the longest known radical chain additions to alkenes are of this polarity type, e.g. the peroxide catalysed anti-Markownikow addition of HBr and the addition of halogenated hydrocarbons such as CCl_4 and $BrCCl_3$. A more recent example is shown in equation 2.36. Here the chain carrying species is the electrophilic radical $\cdot CH(CN)_2$ which adds rapidly to the nucleophilic alkene tetramethylethylene.

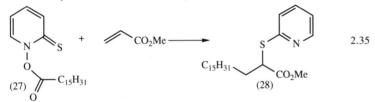

2.36

Other long known intermolecular radical additions to multiple bonds include Meerwein arylation. Here transition metal redox processes catalyse the requisite electron transfer (equations 2.37 and 2.38).

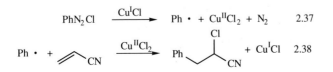

Oxidative coupling of phenols, a reaction long considered to involve radical–radical coupling, may also be a radical addition leading to overall aromatic substitution. The well known oxidation of *p*-cresol (4-methylphenol) in alkaline potassium hexacyanoferrate(III) to 'Pummerer's ketone' (31) could occur by addition of the relatively electron deficient phenoxy radical (29) at its *p*–position to the less hindered electron rich *o* – position of phenoxide ion. A subsequent electron transfer (to Fe^{III}) followed by aromatisation and intramolecular ionic addition to the α,β–unsaturated ketone gives (31), equation (2.39).

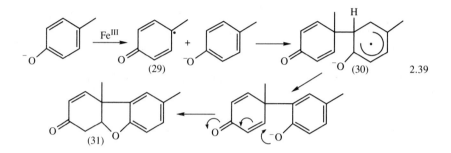

2.39

This mechanism provides an explanation for the well known preference for *o,o-*, *o,p-*, or *p,p*-coupling found in phenolic oxidations in terms of stabilisation of the ketyl radical intermediates, e.g. (30). In general, however, intermolecular homolytic aromatic substitution is not a very useful process synthetically owing to the absence of strong substituent directing effects.

Another important type of intermolecular radical addition to multiple bonds is found for suitable allylically substituted alkenes. Here the basic mode of reaction is an addition–elimination leading to overall allylic substitution, often designated S_H2' by analogy with the corresponding nucleophilic reaction (equation 2.40).

$$R \cdot + \quad \overset{X}{\diagdown} \longrightarrow R \diagup\diagdown X \longrightarrow R \diagup\diagdown + X \cdot \qquad 2.40$$

Typical examples are shown in equations 2.41 and 2.42. The former involves a radical chain sequence with Bu$_3$Sn· as chain carrier, generating an alkyl radical from (32) which then adds to (33). In the latter (equation 2.42), nucleophilic alkyl radicals derived from (34) are trapped by electron deficient reagent (35); Me$_3$CS· radical is then eliminated and acts as chain carrier.

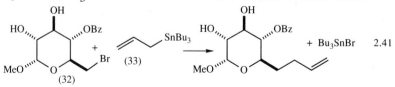

2.41

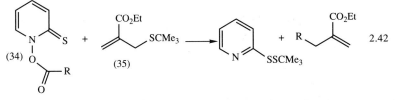

In a similar way, allene transfer via radical capture by the propargyl stannane (36) has been accomplished (equation 2.43).

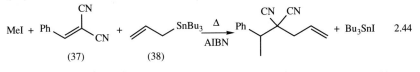

An interesting multiple addition process is shown in equation 2.44 which illustrates a number of points about radical reactions in synthesis. The good yield obtained is a consequence of the preferential addition of the nucleophilic methyl radical to the electron deficient alkene (37), followed by addition of the electrophilic radical generated thereby to the electron rich allyl stannane (38).

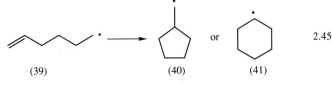

Intramolecular radical addition to multiple bonds

The intramolecular analogues of the intermolecular additions just described are particularly favoured, being subject to fewer limitations in terms of the required substitution pattern of the multiple bond, etc. This is a consequence of the well known proximity effect which leads to more positive entropies of activation than for the intermolecular analogue. In consequence, rates of such exothermic cyclisations, where a σ−bond is formed at the expense of a π−bond, are fast and they can compete efficiently with bimolecular chain transfer steps.

An important prototype is the cyclisation of the hex-5-enyl radical (39) to the cyclopentyl methyl radical (40), which occurs with a rate constant of about 10^6s^{-1} at 20°C (equation 2.45). Noteworthy is the preferred 5-*exo* mode of cyclisation to give (40) compared to the alternative 6-*endo* one giving cyclohexyl radical (41). The latter would be expected to be the more stable, being a secondary radical and having an unstrained cyclohexane ring. This synthetically important result is explained in terms of an early transition state having little product character. Thus stereoelectronic factors govern the cyclisation and optimum overlap between the singly occupied molecular orbital of the radical and the π*-orbital of the double bond is achieved for the five-membered ring transition state.

5-*exo*cyclisation

6-*endo*cyclisation

Typical results for the reduction of hex-5-enyl bromide with tributyltin hydride at 60°C are given in equation 2.46. The preference for five ring closure is clearly shown and the dependence of the cyclic/acyclic product ratio on the tributyltin hydride concentration is as expected for a competition between the unimolecular cyclisation and bimolecular chain transfer steps.

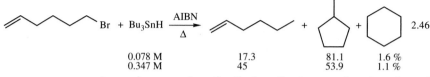

0.078 M	17.3	81.1	1.6 %	
0.347 M	45	53.9	1.1 %	

A very large number of radical cyclisation reactions have been described in the past ten years or so. They differ in substrate type, mode of radical generation, and fate of the cyclised radical. We can do no more than mention a few of these to give some idea of the scope of this method for ring formation.

Typical simple illustrations are shown in equations 2.47 and 2.48. In the former, the radical precursor (42) is obtained by electrophilic acetoxymercuration of the non-conjugated double bond of the starting material and the radical is generated by hydride reduction of the mercurial. The second case (equation 2.48) involves the now conventional tributyltin hydride/alkyl halide radical chain chemistry. In both cases the product formed has the five-membered ring *cis*-fused to the other ring. This is another consequence of the stereoelectronic requirements of radical cyclisations.

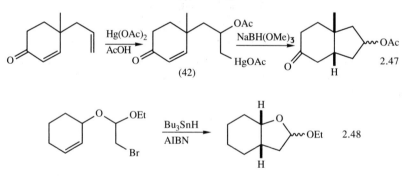

Intramolecular alkyl radical trapping by an electrophilic double bond has also been utilised for the preparation of large ring lactones, e.g. equation 2.49.

The endocyclic mode of ring closure seems to characterise such macrocyclisations

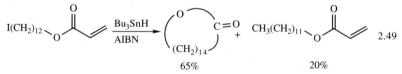

A variety of other ways of forming the required radicals are illustrated by equations 2.50–2.53.

Note that this is a five-membered thiohydroxamic ester, cf. equation 2.34

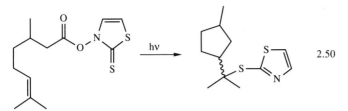

2.50

In equation 2.50 N–O cleavage of the thiohydroxamate has been followed by decarboxylation to give a primary radical which has then undergone cyclisation. A chain sequence is then set up by attack of the tertiary radical so produced on the thiocarbonyl group of another molecule of starting material.

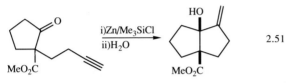

2.51

A trimethylsilylated ketyl radical which undergoes intramolecular addition to the triple bond is presumably involved in the reaction shown in equation 2.51.

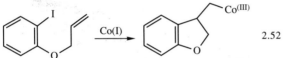

2.52

Here the abbreviation Co(I) refers to a nucleophilic Co^I reagent derived by reduction of the Co^{III} species, BrCo^{III}(salen)PPh$_3$, having the structure shown. Presumably the initial radical is formed by electron transfer from Co^I to the aryl iodide with loss of iodide ion; cyclisation followed by trapping with Co^{II} leads to the cyclised Co^{III} compound, which can be isolated.

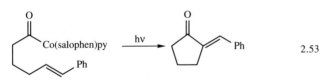

2.53

BrCo^{III}(salen)PPh$_3$

Another example of radical generation via cobalt chemistry is shown in equation 2.53. Here the CO–Co homolysis is followed by 5-*exo* cyclisation, trapping by Co, and elimination to give the unsaturated ketone shown. The salophen ligand in this example is closely related to the salen ligand involved in the previous case. The two examples together show how functionality is maintained in the cyclisation product in contrast to the trialkyltin hydride reductions of unsaturated halides.

In suitably substituted precursors, where the first formed hex-5-enyl radical is stabilised, reversible intramolecular addition–elimination is feasible thereby affording products derived from the thermodynamically more stable six-membered ring. The cyclisation shown in equation 2.54 is considered to be an example. Hydrogen abstraction from starting material to give radical

(44), followed by kinetically favoured but reversible 5-*exo*cyclisation to (43), allows the more stable 6-*endo*cyclised radical (45) to be formed. Chain transfer by hydrogen abstraction from starting material gives the product shown.

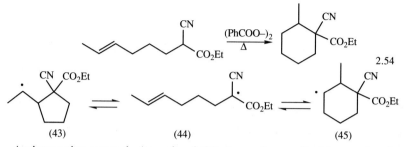

2.54

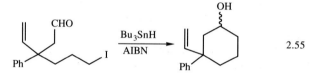

(43)   (44)   (45)

An interesting example (equation 2.55) shows the result of intramolecular competition between 5-*exo*cyclisation onto C=C and 6-*exo* addition to C=O. It is likely that the predominance of the latter process is the result of reversible addition to both multiple bonds but faster capture by the alkoxy radical of hydrogen from tributyltin hydride, another consequence of polarity effects.

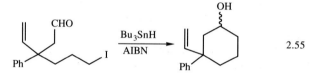

2.55

In favourable cases, the cyclised radical obtained by intramolecular addition can be captured by intermolecular reaction with a suitable trapping agent. Some examples, in which the cyclised radical is directed in various ways according to the trapping species present, are shown (equation 2.56). They include simple addition leading to compound (46); addition, followed by Bu$_3$Sn· elimination giving the enone (47); and addition to an isonitrile followed by Bu^t radical elimination with formation of the cyanide (48). These compounds have been used as precursors in prostaglandin synthesis.

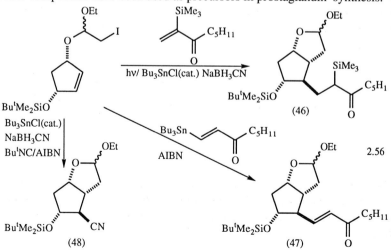

2.56

Double, so-called 'tandem', cyclisations have been imaginatively used in the preparation of polycyclic compounds. A good example is the synthesis of the sesquiterpene hydrocarbon capnellene (equation 2.57).

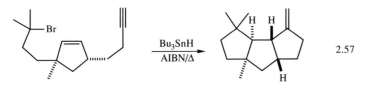

2.57

Note again in equations 2.56 and 2.57, the stereoelectronic preference for radical cyclisation with formation of *cis*-fused five-membered rings

Functional group transformations

Although we have so far concentrated on carbon–carbon bond forming reactions there are a number of useful functional group transformations, particularly reductions, that can be conveniently carried out by free radical chemistry.

The conversion of alkylmercurials to alkanes by sodium borohydride and the reduction of alkyl halides to alkanes by tributyltin hydride have already been described.

Useful methods for deoxygenation of alcohols via xanthates (49), or other thiocarbonyl derivatives, using tributyltin hydride, have been developed (equation 2.58). A chain mechanism along the lines shown is likely. The reaction is general for the xanthates from secondary alcohols at or below toluene reflux temperature and for those from primary alcohols under more vigourous conditions.

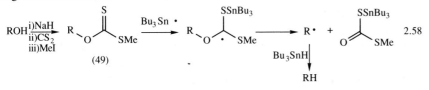

2.58

Because of problems in preparing the appropriate thiocarbonyl intermediates, tertiary alcohols are best deoxygenated using Barton's thiohydroxamic ester chemistry (equation 2.34) via their half oxalates (equation 2.59).

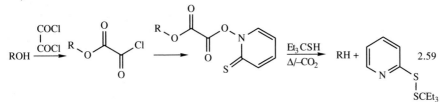

2.59

Radical deoxygenations of the above types in general give satisfactory yields except for systems where good radical leaving groups are β to the radical centre, when elimination leading to alkene formation supervenes (equation 2.60).

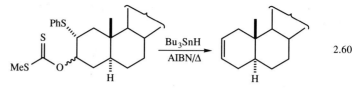

2.60

Tributyltin hydride chemistry can also be used to reduce tertiary and activated secondary nitrocompounds (e.g. equation 2.61). Even primary amines have been converted to the corresponding R–H compounds via the intermediacy of isonitriles (equation 2.62).

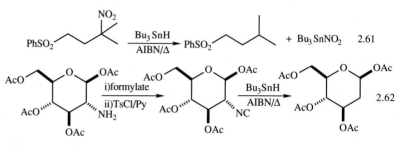

2.61

2.62

There are also a number of useful transformations of carboxylic acids, based on their thiohydroxamic esters, which are exemplified in equation 2.63. Thus RCO_2H is readily converted to RH, ROOH (and by reduction ROH), and RBr, the latter reaction being a modern variant of the Hunsdiecker reaction.

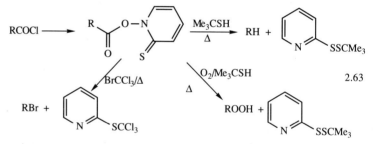

2.63

Functionalisation of unactivated C–H

A number of free radical reactions which are particularly useful from a preparative point of view involve intramolecular hydrogen abstraction from a specific site governed by proximity effects.

One group of such reactions involves alkoxy radicals (cf. equation 2.12) where the conversion of C–H to O–H is exothermic, and the site of hydrogen abstraction is dominated by a strong requirement for a six-membered ring transition state.

Many applications to steroid chemistry are known where the rigid geometry of the steroid skeleton imparts close proximity betwen the alkoxy radical and an angular methyl group (e.g. equation 2.64). A chain mechanism is likely, involving H-abstraction from the angular methyl group by the alkoxy radical, cf. (51), followed by chlorine abstraction from another molecule of the starting hypochlorite (50), etc. As in the example shown, the chloroalcohol formed can often be cyclised on base treatment to form a tetrahydrofuran.

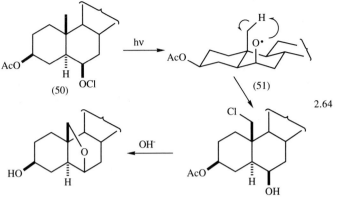

(50)

(51)

2.64

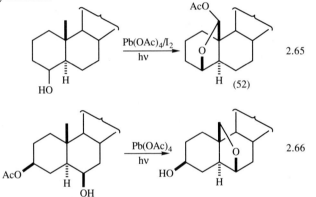

Related reactions are shown in equations 2.65 and 2.66. In the former an alkyl hypoiodite is generated *in situ* and converted, as before, to an iodoalcohol which undergoes further oxidation to the masked aldehyde (52). In the latter, (equation 2.66), the first formed hydroxy lead derivative suffers *in situ* cyclisation.

$$\xrightarrow[hv]{Pb(OAc)_4/I_2}$$

2.65

(52)

$$\xrightarrow[hv]{Pb(OAc)_4}$$

2.66

An important reaction of this type is the Barton reaction of nitrite esters, exemplified by the key step in the synthesis of aldosterone (equation 2.67). A non-chain process is postulated.

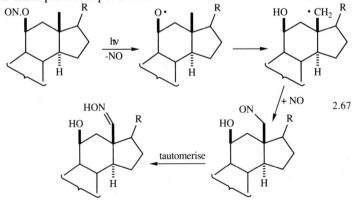

$$\xrightarrow[-NO]{hv}$$

+ NO

2.67

tautomerise

Closely related is the Hofmann–Löffler–Freytag reaction of protonated chloramines which is useful as a general synthesis of pyrrolidines (e.g. equation 2.68). Again the key hydrogen abstraction step in the chain reaction is likely to involve a six-membered transition state.

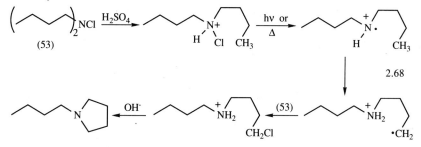

2.68

More recently Breslow has achieved 'remote functionalisation' of C–H groups in steroids by attaching suitable appendages containing groupings which will temporarily trap a halogen atom and thereby direct H abstraction to a precise position (equation 2.69). In both instances a 'radical relay' chain mechanism is proposed involving intermediates of the type (54) and (55).

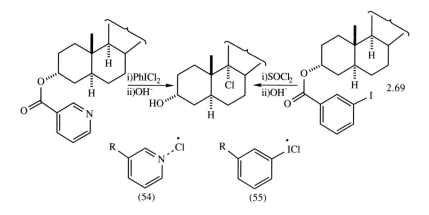

2.69

(54)   (55)

Problems

1. Rationalise the differing behaviour of cyclooocta-1,5-diene in the radical addition reactions shown.

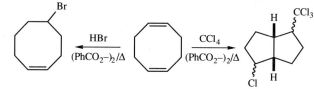

2. Account for the products in the following reaction and explain the observation that the ratio of the products changes significantly when $PhCH_2SH$ is present.

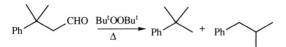

3. Assign a structure to the product of the following transformation:

4. Provide explanations for the following reactions:

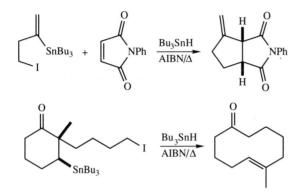

Further reading

For information on the basic physical-organic chemistry of radicals the two volume compendium *Free Radicals,* ed. J. K. Kochi, J.Wiley, New York, 1973, is still very useful.

Radicals in Organic Synthesis: Formation of Carbon–Carbon Bonds, B. Giese, Pergamon Press, Oxford, 1986.

Best Synthetic Methods, Free Radical Chain Reactions in Organic Synthesis, W. B. Motherwell and D. Crich, Academic Press, London, 1991.

The Design and Application of Free Radical Chain Reactions in Organic Synthesis, D. P. Curran, *Synthesis,* 1988, 417 and 489.

Radical Reactions in Organic Synthesis, M. Ramaiah, *Tetrahedron,* 1987, **43**,3541.

3 Carbenes

Carbenes are neutral, divalent, highly reactive carbon intermediates, generally written as $R_2C:$, the two dots indicating that the carbon has two non-bonding electrons. The simplest carbene, $H_2C:$, is usually called methylene, the term being first introduced during the nineteenth century. Indeed before the quadrivalency of carbon was firmly established many early chemists believed that methylene would be a stable entity, and various experiments were attempted for its generation. The word carbene was apparently conceived by Woodward, Doering and Winstein and was first introduced at a meeting of the American Chemical Society in 1951. Nowadays the term is universally used for divalent carbon species which are simply named as substituted derivatives of carbene; for example PhCH is phenylcarbene and $Cl_2C:$ is dichloro-carbene. The exceptions to this are carbenes in which the divalent carbon is part of a ring or is doubly bonded. These are named using the *-ylidene* suffix; for example cyclohexylidene (as shown) and $H_2C=C:$ is vinylidene. The latter structure, containing a doubly bonded divalent carbon, reminds us that carbon monoxide as well as isonitriles can also be written as carbenes, $O=C:$ and $RN=C:$ respectively, although they are probably more accurately represented by their dipolar resonance forms, e.g. $^+O\equiv C^-$.

Cyclohexylidene

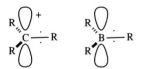

Other species with a vacant *p*-orbital: carbocations and boranes

3.1 Structure and reactivity

Carbenes are usually thought of as being sp^2-hybridised. There are two bonding electrons (both in sp^2-orbitals) and two non-bonding electrons associated with the carbon. Since the non-bonding electrons can have their spins paired or parallel, there is the possibility of two electronic arrangements or *spin states*. The *singlet* carbene (spin *multiplicity* = 1) has the spins of its non-bonding electrons paired. This electron pair is in an sp^2-orbital leaving a vacant *p*-orbital. In *triplet* carbenes, however, the two non-bonding electrons have parallel spins, and both the sp^2- and *p*-orbitals contain one electron each. Hence triplet carbenes (multiplicity = 3) with their unpaired electrons exhibit the properties of diradicals, and, under the right conditions, can be detected by ESR spectroscopy, for example.

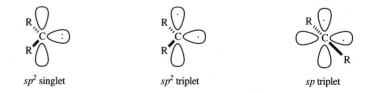

Electronic structures of $R_2C:$ showing sp^2 singlet and triplet, and sp triplet spin states

If carbenes were linear the two *sp*-orbitals would be degenerate and therefore *Hund's Rule* would predict a triplet ground state. In general carbenes are not linear but nevertheless the triplet is often the ground state, although the energy difference between singlet and triplet states is usually small (estimated to be 32–42 kJmol^{-1} for methylene). However the nature of the substituents has an important effect on the electronic properties of carbenes. For example, as the carbene substituents R become better π-donors, then the ground state changes from triplet to singlet. Thus dichlorocarbene is a ground state singlet because electron donation from chlorine into the vacant *p*-orbital on carbon stabilises the singlet state through dipolar resonance structures, although because of the relative electronegativities there is a strong σ-polarisation in the opposite direction. There can be other reasons for the singlet spin state being lower in energy than the triplet. Cycloheptatrienylidene may be such a carbene since in the singlet, provided the molecule is planar, the vacant *p*-orbital can overlap with the rest of the π-system, making a total of six delocalised π-electrons and hence a measure of stabilisation, although the cyclic allene structure, cyclohepta-1,2,4,6-tetraene, may be more stable.

Cycloheptatrienylidene

Stabilisation of singlet spin state of dichlorocarbene by π-donation

The nature of the substituents also affects the chemical reactivity of carbene intermediates. Carbenes are electron deficient intermediates, the carbon atom having only six electrons in its outer shell. Therefore in general carbenes are highly electrophilic in their reactions. Not surprisingly the more electron withdrawing the substituents, the more strongly electrophilic the carbene. Thus difluorocarbene is a more electrophilic intermediate than dichlorocarbene. However if very strong π-donor substituents such as amino groups are present then the carbene may be nucleophilic in its reactions.

Diaminocarbenes – nucleophilic singlets due to strong π-donation by substituents

Until fairly recently the high reactivity and short life times of reactive intermediates had precluded any direct measurements on the species. However the advent of flash photolysis techniques, particularly those employing lasers, has enabled chemists to measure the UV spectra of transient species and hence gain important information about their structure. Matrix isolation methods in which the intermediate is generated within an inert frozen matrix, often argon, at temperatures between 4 and 77K, have also allowed the direct observation of reactive intermediates by IR and ESR spectroscopy. Of particular relevance are the measurements that have been carried out on triplet carbenes, which since they behave as diradicals offer the possibility of

A crystalline carbene!
R = 1-adamantyl

detection by ESR spectroscopy. Thus diazo compounds, the most common carbene precursors (q.v.), can be frozen in an argon matrix and irradiated with UV light within the probe of an ESR spectrometer to generate the carbene, which provided that it has a triplet ground state may have sufficient life-time to give rise to an observable ESR signal. Analysis of such spectra provides details of the carbene structure such as the bond angle between the two substituents; for example triplet diphenylcarbene has a bond angle of about 142°, whereas the much more hindered di(anthracen-9-yl)carbene is almost linear. The best estimate for the bond angle in triplet methylene itself is about 136°; most singlet carbenes are estimated (by various means) to have bond angles of about 102°. Carbenes can be stabilised by steric and/or electronic effects, and two dramatic illustrations of this are the very recent reports of stable carbenes. French workers have prepared $(^iPr_2N)_2P\text{-}C\text{-}SiMe_3$, a distillable red oil, and in 1991 a US group reported the first crystalline carbene!! 1,3-Di-1-adamantylimidazol-2-ylidene is a colourless solid, m.p. 240°C, which is stable in the absence of oxygen. The bond angle at the carbene carbon as determined by X-ray crystallography is 102°.

3.2 Generation of carbenes

Most methods of generating carbenes are based on elimination or fragmentation reactions, and usually involve the breaking of rather weak bonds and the formation of a small thermodynamically stable byproduct such as dinitrogen. The commonly used methods are summarised on the next page.

Diazo compounds

Diazo compounds, known for over a century since the first preparation of ethyl diazoacetate by Curtius in 1883, are probably the most widely used carbene precursors. They possess a 1,3-dipolar structure and are readily prepared from a variety of precursors, decomposing to carbenes with loss of nitrogen under thermal or photochemical conditions or under the influence of certain transition metal catalysts. The stability of diazo compounds is very much influenced by the substituents present. Simple diazoalkanes such as diazomethane tend to be rather unstable and because of their toxic and explosive nature must be handled with great care. They are also rapidly decomposed by acids by a mechanism which involves protonation on carbon followed by loss of nitrogen. Diazo carbonyl compounds on the other hand are much more stable towards heat and acid, as are other diazo compounds in which the diazo 1,3-dipole can be resonance stabilised by electron-withdrawing substituents on carbon. In addition, such compounds are easily prepared by the *diazo transfer* reaction using arenesulphonyl azides.

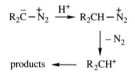

Acid decomposition of diazo compounds

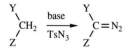

Diazo transfer reaction (Y and Z must be electron withdrawing groups)

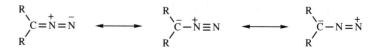

1,3-Dipolar nature of the diazo group indicating possibility for resonance stabilisation

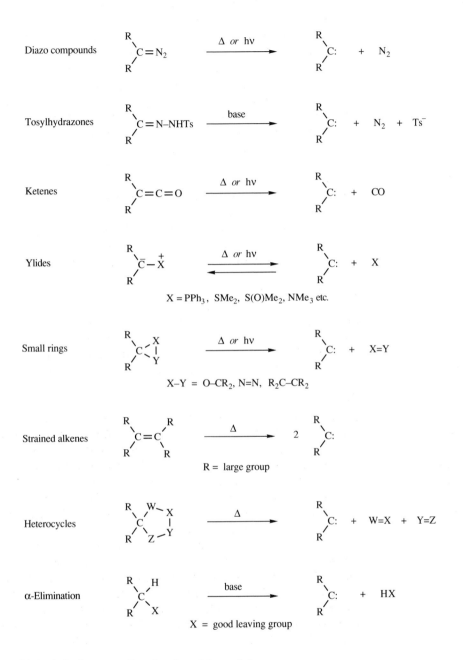

Methods for the generation of carbene intermediates

Transient rhodium carbenoid
intermediate

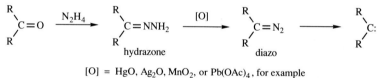

Stable chromium carbene complex

Unless the diazo compound is particularly stable, thermal generation of carbenes is usually preferred since the photochemical method produces intermediates with much higher energy, leading to indiscriminate reactions. However the thermal decomposition of many diazo compounds is markedly catalysed by the presence of certain transition metal salts, a fact that was first noted over 80 years ago. Originally copper powder or copper(II) salts were used for such purposes, but nowadays rhodium(II) carboxylates such as rhodium(II) acetate are usually the preferred catalysts. In the presence of such catalysts diazo carbonyl compounds readily decompose at room temperature in dichloromethane or in boiling benzene to give carbene-type products. The intermediates in these reactions are not free carbenes but transition metal complexes often referred to as *metallocarbenes* or *carbenoids* and usually represented by the structure shown (L = ligand) containing a formal metal–carbon double bond. Such transition metal carbenes possess the same electron deficient nature as free carbenes and undergo the same types of reaction, and although not isolable bear some structural resemblance to stable metal–carbene complexes of, for example, tungsten and chromium. The recent development of chiral copper and rhodium catalysts with the potential for asymmetric synthesis has further increased the usefulness of the catalytic decomposition of diazo compounds.

Tosylhydrazones – the Bamford–Stevens reaction

In cases where the diazo compound is somewhat unstable, low molecular weight diazoalkanes for example, then it often better to use a diazo precursor. The simplest of these are hydrazones readily prepared from ketones and converted into diazo compounds using oxidants such as mercury(II) oxide, silver(I) oxide, manganese(IV) oxide, or lead(IV) acetate. The diazo compound may be isolated or can be taken straight through to carbene derived products. The method finds application in the synthesis of alkynes from 1,2-diketones via the bis-hydrazones, although non-carbene mechanisms can be written for this transformation.

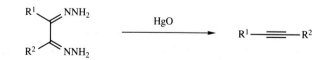

[O] = HgO, Ag$_2$O, MnO$_2$, or Pb(OAc)$_4$, for example

Generation of carbenes from hydrazones

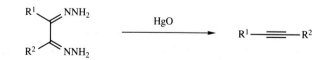

Preparation of alkynes by oxidation of bis-hydrazones

Much more widely used carbene precursors are tosylhydrazones which are readily prepared from aldehydes or ketones by reaction with 4-

toluenesulphonyl hydrazide. On treatment with base such as sodium methoxide or sodium hydride the acidic NH proton is removed and the tosylhydrazone sodium salt can be isolated. On heating or on irradiation the salt decomposes with loss of sodium 4-toluenesulphinate (NaTs) and nitrogen to give the carbene. The whole process is known as the Bamford–Stevens reaction, and is quite general for a range of carbenes. Under certain circumstances the intermediate diazo compounds may be isolated although this is rare.

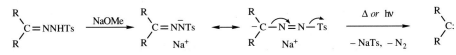

The Bamford–Stevens reaction; carbenes from tosylhydrazones

Ketenes

Ketenes, like diazo compounds, can eliminate a stable molecule (CO) on thermolysis or photolysis to generate carbenes. The reaction, which is illustrated in general (right), is not widely used since ketenes are not readily available precursors and tend to polymerise under the reaction conditions.

$$R_2C=C=O \xrightarrow[-CO]{\Delta \ or \ h\nu} R_2C:$$

Ylides

Both phosphorus and sulphur ylides are well known reagents in synthetic organic chemistry and react with carbonyl compounds to give alkenes (the Wittig reaction) and epoxides respectively. Ylides also react with electrophilic alkenes to give cyclopropanes, the formal products of carbene addition to the double bond. However the mechanism involves conjugate nucleophilic addition of the ylide followed by elimination to form the three-membered ring rather than a carbene intermediate. Nevertheless such ylides are often known as carbene transfer reagents.

$$\overset{+}{X} - \overset{-}{C}H_2$$

Phosphonium, sulphonium and sulphoxonium ylides, e.g. X = Ph$_3$P, Me$_2$S and Me$_2$SO respectively

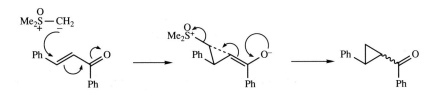

Cyclopropane formation from ylides and electrophilic alkenes (non-carbene mechanism)

However under photochemical conditions phosphorus, sulphur and nitrogen ylides often give carbene type products in their reactions, although the intermediacy of free carbenes has not been proven beyond doubt. It should also be noted that, as indicated in the general scheme on page 29 the reaction is reversible since electrophilic carbenes readily react with nucleophiles to give ylides (see later).

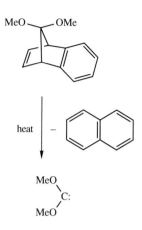

Generation of dimethoxycarbene by retro-1,4-cycloaddition

Small rings

Three-membered rings, which have a high ground state energy because of steric strain, will often decompose to give carbene intermediates simply on heating or irradiation, provided that a thermodynamically stable fragment is extruded. Examples involving epoxides and cyclopropanes are shown below. Epoxides are useful but rarely used precursors to arylcarbenes, whereas the decomposition of cyclopropanes is the reverse of their formation from a carbene and an alkene, and also of limited use. Related to this formal retro-1,2-cycloaddition is the retro-1,4-cycloaddition of benzonorbornadienes which decompose on heating to generate carbenes with the elimination of naphthalene. The method is commonly used as a source of dimethoxycarbene and is also widely used for the generation of silylene and germylene intermediates.

Generation of carbenes from epoxides and cyclopropanes

The most useful carbene precursors of the small ring type are diazirines. Diazirines are cyclic isomers of diazo compounds and it is interesting to note that an early structure proposed for diazomethane itself was the three-membered diazirine, although it was many years after this before the first genuine diazirine was isolated. Surprisingly perhaps diazirines are not particularly hazardous to handle despite their obvious desire to lose molecular nitrogen. They are readily prepared from ketones by reaction with ammonia and chloramine followed by oxidation of the resulting diaziridine. The method is quite widely used, particularly for halocarbenes and for mechanistic studies.

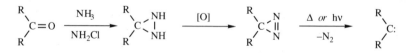

Preparation and decomposition of diazirines

Strained alkenes

The cleavage of a carbon–carbon double bond to give two carbenes is thermodynamically highly unfavourable, the decomposition of ethene into two molecules of methylene, for example, is reckoned to be endothermic by about 800 kJmol^{-1}. However if the alkene is extremely sterically hindered then the π-bond is weakened due to the considerably reduced *p–p* overlap and distortion from planarity thereby raising the ground state energy, and therefore dissociation to carbenes on heating is not prohibitively costly in energy. The well-known example of this process is the reversible dissociation of tetranaphth-1-ylethene into two molecules of bis(naphth-1-yl) carbene on heating to about 250°C.

R = naphth-1-yl

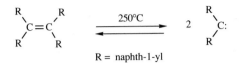

R = naphth-1-yl

Thermal dissociation of tetranaphth-1-ylethene

Related to the dissociation of sterically crowded alkenes to carbenes is their rearrangement on irradiation. For example, 1,1-di-*tert*-butylethenes rearrange by hydrogen shift to generate alkylcarbenes in the reverse of a common carbene rearrangement pathway, the driving force presumably being the relief of strain in going from an sp^2- to an sp^3-carbon which bears two bulky groups. On strong heating, usually in the vapour phase, alkynes can also rearrange by hydrogen shift to give vinylidenes. The reaction has been exploited as a route to cyclopentenones since the vinylidenes readily undergo intramolecular C–H insertion reactions.

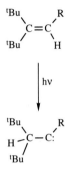

Rearrangement of crowded alkene

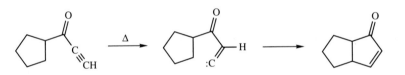

Generation of a vinylidene from an alkyne and subsequent intramolecular C–H insertion

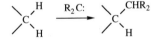

C–H insertion (see Section 3.4)

Heterocycles

Many five-membered ring heterocyclic compounds decompose to give carbenes on heating or irradiation with the extrusion of thermodynamically stable fragments. The decomposition of 1,5-dihydro-1,3,4-oxadiazoles shown below occurs readily at about 80°C with loss of nitrogen followed by the carbonyl fragment to give carbenes. In certain cases the intermediate carbonyl ylide, $R_2C=O^+-C^-Me(OMe)$ in the example cited, may be trapped in 1,3-dipolar cycloaddition reactions. Another example is provided by 1,3-dioxolan-2-ones which fragment with loss of carbon dioxide and a carbonyl compound. Five-membered heteroaromatic rings can also fragment to carbenes although the aromatic stabilisation usually means that the reaction must be carried out at quite high temperature or photochemically.

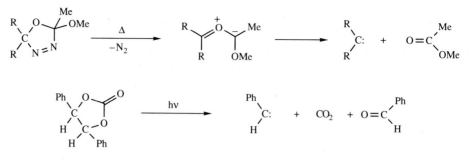

Fragmentation of five-membered heterocycles to carbenes

α-Elimination

HCCl₃ $\xrightarrow[\text{DO}^-]{\text{D}_2\text{O}}$ DCCl₃

Historically this is one of the most important routes to carbenes, since it was the investigation into the mechanism of the hydrolysis of chloroform under basic conditions, with the formation of carbon monoxide and formate, by Hine and co-workers in the early 1950s that initiated the modern era of carbene chemistry, although the idea that dichlorocarbene might be an intermediate in the reaction was first expressed by Geuther ninety years earlier in 1862. The carbene mechanism, which involves rapid formation of the stabilised trichloromethyl anion followed by rate determining loss of chloride by α-elimination and subsequent trapping of the carbene by the aqueous solvent, was proven by detailed kinetic studies of the reaction.

$$Cl_3CH \underset{H_2O}{\overset{^-OH}{\rightleftharpoons}} Cl_3C^- \xrightarrow[-Cl^-]{\text{slow}} Cl_2C: \xrightarrow{H_2O} H_2\overset{+}{O}-\overset{-}{C}Cl_2 \xrightarrow[-Cl^-]{-H^+} \underset{HO}{\overset{Cl}{>}}C: \xrightarrow[-Cl^-]{-H^+} CO \xrightarrow[\text{slow}]{^-OH} HCO_2^-$$

Basic hydrolysis of chloroform showing involvement of dichlorocarbene

Cl₃CH

| 50% aq. NaOH
| PhCH₂NEt₃Cl
↓

Cl₂C:

Phase transfer generation of dichlorocarbene

The reaction has been extended to other halomethanes and is a general route to dihalocarbenes. The ease of the elimination depends on the halide leaving group and follows the usual order: I > Br > Cl >> F. Thus CHClF₂ gives exclusively difluorocarbene, CHBr₂F gives bromo(fluoro)carbene and so on. The reactions are best carried out in an organic solvent using a strong non-nucleophilic base such as potassium *tert*-butoxide, although nowadays the use of phase transfer catalysis has become the route of choice for the generation of dichlorocarbene. The method involves taking a two-phase mixture of chloroform (which also contains the substrate for the carbene such as an alkene) and 50% aqueous sodium hydroxide solution, and adding a quaternary ammonium salt, e.g. PhCH₂NEt₃Cl, which acts as the phase-transfer catalyst to 'transport' the hydroxide base into the organic phase. Although the reaction is run in the presence of water, the carbene is generated and reacts in the organic phase.

If the strongly basic conditions required to generate dihalocarbenes by α-elimination are likely to cause unwanted side reactions, then a route which avoids the use of external base must be used. The sodium salt of trichloroacetic or tribromoacetic acid is a convenient precursor to dichloro or dibromo carbene since on heating in an aprotic solvent such as 1,2-dimethoxyethane (b.p. 80°C) it readily undergoes decarboxylation followed by loss of sodium chloride (or bromide) from the trihalomethyl anion. A related method involves the use of phenylmercury trihalomethanes, $PhHgCX_3$, often called Seyferth reagents after the chemist who developed their use. On heating in benzene the reagents eliminate PhHgX with the formation of the dihalocarbene.

    Attempts to extend the above α-elimination reactions to the generation of carbenes other than dihalocarbenes have been largely unsuccessful. Thus dichloromethane is not a good source of chlorocarbene under the usual basic conditions, although reasonable yields of carbene adducts can be obtained when butyllithium is used as base. However, this and analogous reactions of 1,1-dihalo compounds almost certainly involve an α-halo-organolithium species as the intermediate rather than a free carbene. The mechanism involves metal–halogen exchange to give the organolithium intermediate; these are stable in solution at low temperature and have been studied in detail, for example by ^{13}C NMR spectroscopy thereby leaving little doubt about their structure. On being allowed to warm up in the presence of an appropriate substrate, lithium halide is lost and carbene products are formed. Hence the R_2CXLi intermediates show similar properties to the free carbenes $R_2C:$, and for this reason are often known as *carbenoids* or *lithium carbenoids*. Unfortunately the carbenoid terminology is also used for the intermediates formed in transition metal catalysed decomposition of diazo compounds (see page 30). Related zinc carbenoids are almost certainly involved as intermediates in the Simmons–Smith cyclopropanation of alkenes which uses diiodomethane and zinc metal in ether.

$X_3CCO_2^-\ Na^+$

$PhHgCX_3$

Sodium trihaloacetates and the Seyferth reagents: thermal sources of dihalocarbenes (X = Cl, Br)

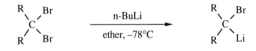

Generation of lithium carbenoids from 1,1-dihalo compounds

The Simmons–Smith reaction

3.3 Cycloaddition reactions of carbenes

Cycloaddition to alkenes

The formation of cyclopropanes by the 1,2-addition of carbenes to alkenes was first reported by Doering in 1954. The process which is probably the most characteristic reaction of carbene intermediates is now widely used as a synthetic route to cyclopropanes. Much of the initial work involved the addition of dibromocarbene, a typical electrophilic singlet carbene, to alkenes to give, it was noted, cyclopropanes in which the alkene stereochemistry was preserved. This result was rationalised in the *Skell hypothesis* (1956) which argued that since a singlet carbene could add to an alkene in a single step process, the two new bonds being formed simultaneously, the addition would be stereospecific. The triplet carbene on the other hand cannot give a cyclopropane in a one-step reaction since this is spin forbidden, and must therefore add in two steps via a diradical intermediate which must undergo spin inversion before the second C–C bond can form. However spin inversion can be relatively slow, and if bond rotation is faster, then stereospecificity is lost. This does not mean that triplet carbenes always add to alkenes in a non-stereospecific manner since spin inversion may be faster than bond rotation.

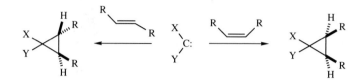

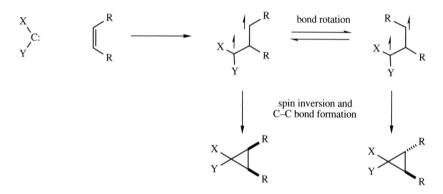

Stereospecific addition of singlet carbenes to alkenes

Non-stereospecific addition of triplet carbenes to alkenes

The addition of singlet carbenes to alkenes is generally regarded as a concerted process, and therefore should obey the rules of orbital symmetry. However it turns out that the direct approach of a carbene to a double bond is forbidden since there are antibonding interactions whichever combination of orbitals one takes. Therefore the carbene adopts a sideways non-linear approach in which bonding interactions between the relevant orbitals are strictly maintained, the dominant interaction usually being HOMO (alkene)–LUMO (carbene), i.e. the carbene 'attacks' as an electrophile with its vacant *p*-orbital, possibly forming a 'loose' π-complex.

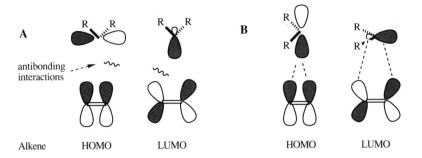

The addition of a carbene to a double bond showing (A) the antibonding interactions for linear approach and (B) the 'allowed' sideways approach

Whatever the exact mechanism, the Skell hypothesis generally holds, and intermediates such as dihalocarbenes which have ground state singlets add stereospecifically to alkenes. The stereochemical outcome of other reactions is often subject to experimental conditions since despite the fact that the carbene may have a triplet ground state, it is usually generated as the singlet. Therefore the stereochemistry of the cyclopropane product depends on how fast the singlet carbene adds to the alkene relative to how fast it decays to its triplet ground state. For example, fluorenylidene generated by photolysis of diazofluorene adds largely stereospecifically to *cis*-but-2-ene since the first formed intermediate is the singlet. However if the reaction is run in the presence of solvents such as dibromomethane or hexafluorobenzene which are good at deactivating the higher energy singlet carbene to its triplet ground state, then not surprisingly the stereospecificity of the addition falls markedly. Conversely if the reaction is conducted in the presence of *triplet quenchers,* substances such as butadiene which selectively remove any triplet carbenes, the addition is again stereospecific.

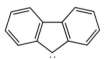

Fluorenylidene

Another mechanistic complication arises if the carbene is generated by thermolysis of a diazo compound. Since the diazo compound is a 1,3-dipole it can undergo cycloaddition to the alkene in its own right *before* it loses nitrogen. The resulting five-membered ring, a pyrazoline, can then lose nitrogen by a diradical mechanism, to give the cyclopropane non-stereospecifically by a mechanism which does not involve a carbene intermediate. The use of transition metal catalysts to decompose the diazo compound avoids this complication.

1,3-Dipole

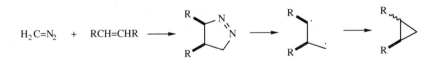

Formation of cyclopropanes by 1,3-dipolar cycloaddition of the diazo compound (non-carbene mechanism)

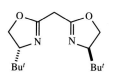

Chiral ligand, L*, for asymmetric cyclopropanation using copper catalysis. The copper , and hence the copper carbenoid intermediate, is in an asymmetric environment

Despite the above mechanistic complications, the inter- and intra-molecular addition of carbenes to alkenes is a widely used route to cyclopropanes, and some examples are shown below. The asymmetric cyclopropanation of styrene to give a mixture of *cis* and *trans* products, both in high enantiomeric excess, using a chiral catalyst system, one of many currently being studied, is particularly noteworthy. Further examples are given in Section 3.7. The exceptions to this general reaction of carbenes are simple alkylcarbenes which rearrange too fast (see later) and nucleophilic carbenes, although these do add to electron deficient alkenes.

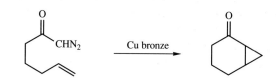

Cyclopropanation of styrene, including asymmetric variant

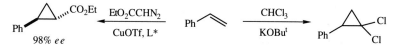

Intramolecular cyclopropanation of an alkene

Dimethylallene – a probe for carbene multiplicity

Cycloaddition to 1,2-dienes (allenes)

Many carbenes add readily to allenes to give alkylidenecyclopropanes in a synthetically useful reaction. Thus allene itself reacts with ethoxycarbonylcarbene generated from ethyl diazoacetate to give the ethyl ester of 2-methylenecyclopropane-1-carboxylic acid. 3-Methylbuta-1,2-diene (dimethylallene) has been proposed as a useful probe for carbene multiplicity since singlets add to the more substituted double bond whereas triplets add to the 1,2-bond. The proposal is certainly borne out by dibromocarbene which gives cleanly 1,1-dibromo-2,2-dimethyl-3-methylene cyclopropane.

Cycloaddition to 1,3-dienes

The 1,4-addition of carbenes to 1,3-dienes to give cyclopentenes is extremely rare, since the 1,2-addition to give a vinylcyclopropane is much more favourable. Unfortunately the situation is complicated by the fact that vinylcyclopropanes can be converted into cyclopentenes on heating although this reaction generally requires high temperatures. In special circumstances direct 1,4-addition reactions are observed, but only in poor yield. Thus dichlorocarbene adds to 1,2-bismethylenecycloheptane, in which the two double bonds are held close to one another, in 0.6% yield.

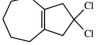

1,4-Addition of dichlorocarbene

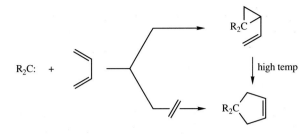

1,2-Addition to 1,3-dienes to give vinylcyclopropanes

Cycloaddition to alkynes

In general alkynes being less nucleophilic than alkenes are less reactive towards carbenes, but nevertheless the reaction is a useful route to cyclopropenes as exemplified by the addition reactions of dibromocarbene to diphenylacetylene and ethoxycarbonylcarbene to phenylacetylene shown below. In the first instance the resulting cyclopropene is readily hydrolysed to the corresponding cyclopropenone, and the second reaction is complicated by competing addition of the carbene as a 1,3-dipole to give a furan.

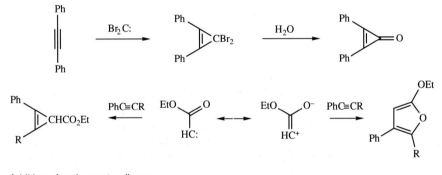

Addition of carbenes to alkynes

Cycloaddition to arenes

The addition of ethoxycarbonylcarbene, generated by thermolysis of ethyl diazoacetate, to benzene is one of the oldest carbene reactions known. It was investigated by Curtius and Buchner in 1885, and in 1896 the initial product was correctly assigned as the norcaradiene, although this readily rearranges to a mixture of cycloheptatrienyl esters by what we would now call an electrocyclic process. The decomposition of diazo compounds in the presence of aromatic rings is catalysed by transition metal salts and an intramolecular version of the reaction has been used as a route to 7–5-bicyclic systems.

EtO_2CCHN_2

Ethyl diazoacetate

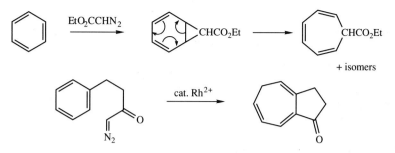

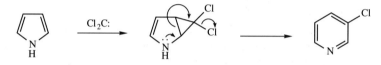

Inter- and intra-molecular addition of carbenes to aromatic rings

Benzene itself is inert to less reactive species like dichlorocarbene although more reactive arenes with greater 'isolated alkene character' such as alkoxynaphthalenes and phenanthrenes react easily. Electron rich five-membered heteroaromatic systems, however, are readily attacked by dihalocarbenes. The reaction is best known for pyrroles and indoles since it constitutes a ring expansion method to give pyridines and quinolines respectively.

Addition of dichlorocarbene to pyrrole

Cycloaddition to other double bonds
The formal cycloaddition of carbene to polarised double bonds such as C=O, C=NR, and C=S is also known. However the reaction almost invariably involves attack of the heteroatom lone pair on the electrophilic carbene and is therefore considered later along with other nucleophiles.

3.4 Insertion reactions of carbenes

Insertion into C–H bonds

Insertion refers to the overall process, rather than a detailed mechanism.

Another characteristic reaction of carbenes is their insertion into single bonds, notably C–H bonds. Provided that it is selective (see below), the insertion reaction is useful in synthesis because it leads to the formation of a new C–C bond. As for cycloaddition to alkenes, two mechanisms can be envisaged: a direct concerted insertion of the singlet, or a stepwise hydrogen abstraction recombination mechanism involving the triplet. The stereochemical consequences of these two mechanisms are that singlet carbenes should insert into C–H bonds with retention of configuration, whereas in triplet reactions which may involve a diradical intermediate, the stereochemical integrity of the substrate may be lost, although there are exceptions to this.

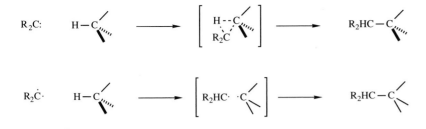

Insertion of singlet and triplet carbenes into a C–H bond

Very reactive carbenes such as methylene insert almost indiscriminately into hydrocarbon C–H bonds. Thus decomposition of diazomethane in pentane or 2,3-dimethylbutane showed that the primary, secondary and tertiary C–H bonds were attacked at random, and statistical ratios of products were formed. Other carbenes are more chemoselective, and ethoxycarbonylcarbene, for example, inserts preferentially into tertiary then secondary over primary C–H bonds. Less reactive carbenes such as dihalocarbenes only insert into more reactive C–H bonds, such as benzylic ones. Insertion into allylic C–H bonds is complicated by competing addition to the double bond. Alkylcarbenes almost invariably react by intramolecular C–H insertion to give cycloalkanes, which is transannular in the case of cycloalkylidenes, or by rearrangement to give alkenes. Further examples of C–H insertion reactions are shown in Section 3.7.

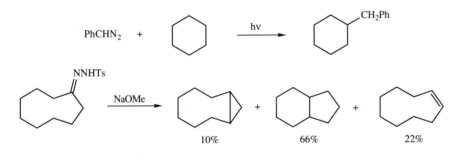

C–H Insertion reactions of carbenes (the alkene in the last example is formed by a hydrogen shift)

Insertion into X–H bonds

Carbenes also insert into other single bonds including C–O, C–N, C–Cl, C–Si, O–H, N–H, S–H, and Si–H, although only the reactions with X–H bonds will be discussed briefly. Carbenes react with alcohols to give the products of O–H insertion although the mechanism probably involves nucleophilic attack by the oxygen rather than a true insertion. Likewise the reaction with amines can lead to the formation of new C–N bonds. Many of these reactions are best carried out via transient transition metal carbenoid intermediates, and some examples of their use are given in Section 3.7.

Required stereochemical
arrangement for 1,2-shift in
carbenes

R = H >> aryl > alkyl

3.5 Rearrangement of carbenes

Carbenes, like other electron deficient intermediates with a vacant *p*-orbital such as carbocations, undergo facile rearrangement in which an atom or group on the adjacent carbon migrates to the electron deficient centre with simultaneous formation of a new C=C bond. Such rearrangements are usually called 1,2-shifts, and often involve migration of a hydrogen atom since the order of migrating ability is: H >> aryl > alkyl. This hydrogen shift, which can also be regarded as an intramolecular insertion of the carbene into the adjacent C–H bond, is analogous to Wagner–Meerwein shifts in carbocations, and proceeds so readily that, as noted previously, intermolecular reactions of simple alkylcarbenes are rarely seen. The mechanism involves overlap of the migrating σ-bond with the vacant *p*-orbital of the carbene, and like many such reactions requires the correct spatial alignment of orbitals with the migrating group coplanar with the vacant orbital. Although the rearrangement is often represented as occurring in the free carbene, in most cases the alternative 'concerted' mechanism in which the migration occurs at the same time as the leaving group (e.g. nitrogen from a diazo compound) is lost cannot be ruled out.

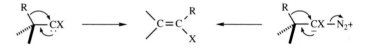

Formation of alkenes by 1,2-shift: carbene and concerted mechanisms

One alkylcarbene rearrangement that deserves special mention is the rearrangement of cyclopropylidenes to allenes, the second of two carbene steps in the so-called carbon insertion reaction, which converts alkenes into allenes, the first being the addition of dibromocarbene to the alkene.

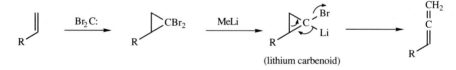

Carbon insertion reaction: conversion of alkenes into allenes

Probably the best known carbene rearrangement is the *Wolff rearrangement* of diazoketones to ketenes. The reaction is often written as a carbene process although there is no conclusive evidence that free carbenes are involved, and it could equally well be a concerted process, the exact mechanism no doubt being dependent on the substituents. The ketene that is formed is not usually isolated but is trapped by an added nucleophile such as water or an alcohol to give an acid or ester as in the Arndt–Eistert homologation of carboxylic acids.

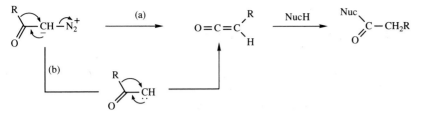

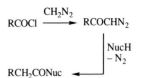

The overall process of the Arndt–Eistert reaction

Wolff rearrangement of diazoketones to ketenes showing concerted (path a) and carbene (path b) mechanisms, and subsequent trapping of the ketene with a nucleophile NucH. If the group R is optically active, it migrates with retention of configuration

Before leaving the topic of carbene rearrangements it should be noted that carbene to carbene rearrangements are also possible. One of the best known is the thermodynamically favoured *Skattebøl rearrangement* of vinylcyclopropylidenes to cyclopentenylidenes which subsequently give cyclopentadienes by 1,2-hydrogen shift. Arylcarbenes also undergo a series of reversible rearrangements when generated in the gas-phase by thermolysis of aryldiazomethanes. Various intermediates, such as the incredibly strained allene, cyclohepta-1,2,4,6-tetraene, have been identified by IR spectroscopy after trapping the species in an argon matrix at low temperature.

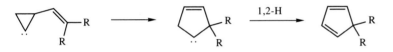

Skattebøl rearrangement of vinylcyclopropylidenes to cyclopentenylidene

Reversible rearrangement of phenylcarbene

3.6 Reactions of carbenes with nucleophiles

Carbenes for the most part being highly electron deficient and hence electrophilic in their reactions, react with nucleophiles of all types. Tertiary amines, phosphines, ethers, sulphides, and sulphoxides all react to give ylides in the reverse of a carbene forming reaction, although the reaction with sulphoxides is complicated by competing deoxygenation. Depending on the substituents present, the ylides may be isolable or may undergo rearrangement. Polarised multiple bonds such as C=O and C≡N also react with carbenes by attack of the heteroatom lone pair to give ylides which undergo further reaction. Some examples of the formation of ylides from carbenes, which are usually best carried out via the corresponding rhodium carbenoids, are shown.

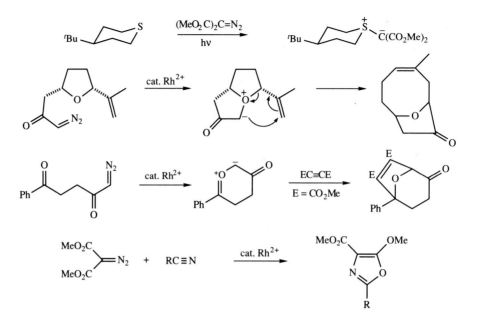

Reaction of nucleophiles (sulphide, ether, carbonyl, nitrile) with carbenes (or rhodium carbenoid)

Carbon nucleophiles such as stabilised anions also react with carbenes, and the Reimer–Tiemann reaction is a well known example of such a process. Phenol is reacted with chloroform and base to give salicaldehyde by initial attack of the phenoxide ion, through carbon, on the electrophilic dichlorocarbene that is generated under the reaction conditions. The resulting adduct rearomatises and the dichloromethyl group is hydrolysed to an aldehyde with assistance from the neighbouring phenoxide ion, possibly via the orthoquinomethide intermediate shown.

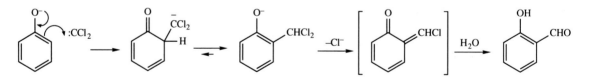

The Reimer–Tiemann reaction of dichlorocarbene with phenoxide ion; the Cl_2C: is generated from $CHCl_3$ as previously described

The formal dimers of carbenes, R_2C:, the alkene $R_2C=CR_2$, are occasional byproducts, probably formed by reaction of one carbene, as an electrophile, with its precursor.

3.7 Uses of carbenes in organic synthesis

Carbene reactions have found wide application in preparative organic chemistry in recent years, particularly in the synthesis of cyclic systems using intramolecular reactions. Some representative syntheses in which a carbene reaction features as the key step are discussed below.

Cyclopropanes

The cyclopropanation of alkenes using carbenes is such a versatile reaction that it is no surprise to find that it has been widely used in the synthesis of cyclopropanes. In the first example presented below, the tricyclic naturally occurring cyclopropane cycloeudesmol is assembled using an intramolecular copper carbenoid addition to a methylene cyclohexane derivative, the substrate diazo compound being readily prepared by a diazo transfer reaction.

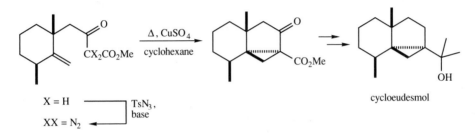

Intramolecular cyclopropanation in the synthesis of cycloeudesmol

The second example involves the preparation of a tetracyclic cyclopropapyrroloindole, an analogue of the mitosene antitumour agents. Again the key step is an intramolecular addition to an alkene, although in this case the reaction probably proceeds through the 1,3-dipolar cycloaddition of the intermediate diazo compound rather than a free carbene.

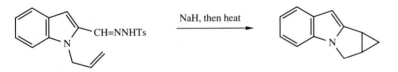

Intramolecular cyclopropanation in the synthesis of a tetracyclic indole

Cyclopentanoid natural products

Cyclopropanes often feature as key intermediates in synthesis, and the following route to the triquinane hirsutene illustrates this. Copper(II) catalysed decomposition of the diazo compound results in formation of a vinyl cyclopropane which on heating undergoes rearrangement to a cyclopentene thereby establishing the tricyclic system in two simple steps. Final reduction and Wittig olefination gives the natural product, hirsutene.

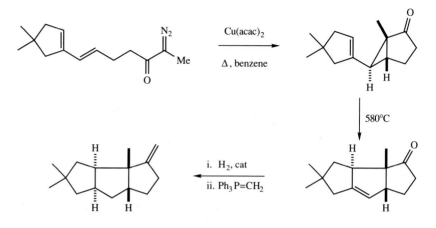

Synthesis of the triquinane hirsutene using intramolecular cyclopropanation

Cyclopentanes are also readily prepared using intramolecular C–H insertions. Most of the reactions involve the rhodium(II) catalysed decomposition of α-diazo-β-keto esters which give cyclopentanones in good yield. Three examples are given below and feature as key steps in routes to α-cuparenone, pentalenolactone E, and an analogue of the ginkolide natural products.

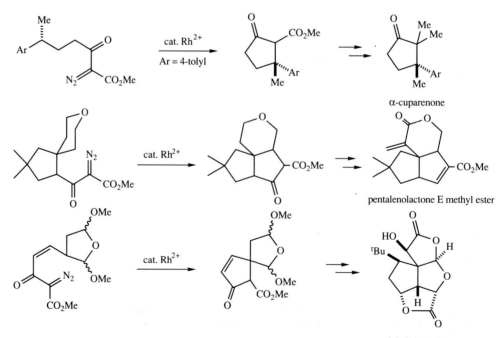

Synthesis of cyclopentanones by intramolecular C–H insertion reactions

β-Lactams

The β-lactam antibiotics, e.g. penicillin, play a key role in the treatment of bacterial infections, and therefore, not surprisingly, organic chemists are always looking for new ways in which to prepare these important compounds. Three routes based on carbenes are shown below. The first involves an intramolecular C–H insertion reaction of a carbene, derived by photolysis of a diazo amide, to form the four-membered ring itself, whereas the second involves a rhodium carbenoid insertion into the N–H bond of an existing β-lactam which allows the fusion of the second ring. This important reaction, which is arguably the most significant carbene reaction of recent years, was originally developed by the Merck company as a key step in the laboratory synthesis of the carbapenam family of antibiotics, and, surprisingly perhaps, is now used in the industrial process. The final example involves a Wolff rearrangement of a diazoketone to effect a ring contraction of a five-membered ring to the β-lactam. Note that it is the C–C bond rather than the C–N bond (which has amide double bond character) which migrates to the electron deficient centre.

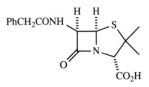

Penicillin G

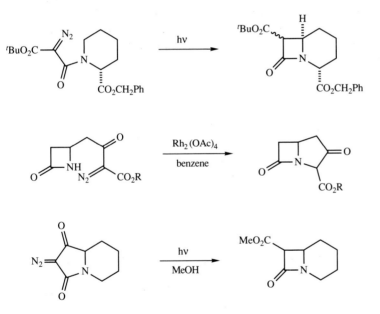

Carbene reactions in the synthesis of β-lactam antibiotics

Heterocycles

The use of carbenes in the synthesis of heterocyclic compounds has been discussed above, and in Section 3.6. Two further examples are presented here and involve an intramolecular O–H insertion reaction and an intramolecular attack on sulphur, followed by 1,2-rearrangement of the resulting ylide. Again both reactions are rhodium catalysed reflecting modern practice.

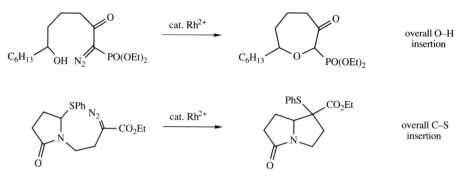

Intramolecular O–H insertion and nucleophilic attack on sulphur

Ring modification reactions

As well as being involved in ring forming reactions, carbenes also feature in useful transformations of existing rings. The ring expansion of cyclic ketones via reaction of their trimethylsilyl enol ethers with methyl-(chloro)carbene is a useful method, and the fragmentation of carbenes derived from α,β-epoxyketones is a general route to acetylenes. The latter reaction, often known as the *Eschenmoser fragmentation,* employs tosylhydrazones or aziridinylimines as the carbene precursor and works particularly well for cyclic alkynes.

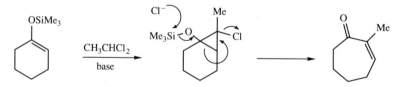

Carbene mediated ring expansion reaction

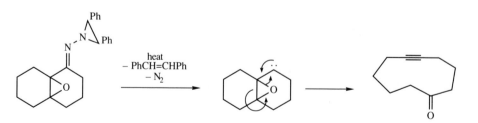

The Eschenmoser fragmentation reaction

Problems

1. Predict the product of the following carbene reaction:

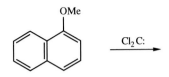

2. Write a mechanism for the following transformation:

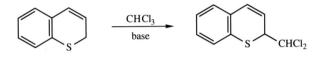

3. Rationalise the difference in behaviour of the following diazoketone under the two sets of conditions shown:

4. Predict the product(s) of the following rhodium carbene reaction:

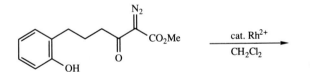

5. Suggest a mechanism for the following reaction:

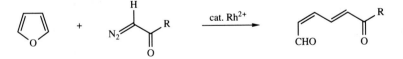

Further reading

For an amusing account of early carbene research by one D. Duck, the reader is referred to *Organic Chemistry*, R. T. Morrison and R. N. Boyd, 5th edn, Allyn and Bacon, Boston, 1987, p. 466.

Carbene chemistry, W. Kirmse, 2nd edn, Academic Press, New York, 1971.

Carbenes, ed. M. Jones and R. A. Moss, Wiley, New York, Vol. I, 1973 and Vol. II, 1975.

Lithium halocarbenoids, H. Siegel, *Top. Curr. Chem.,* 1982, **106**, 55.

Reactive molecules: the neutral reactive intermediates in organic chemistry, C. Wentrup, Wiley, New York, 1984.

Catalytic methods for metal carbene transformations, M. P. Doyle, *Chem. Rev.,* 1986, **86**, 919.

Diazo compounds: properties and synthesis, M. Regitz and G. Maas, Academic Press, Orlando, Florida, 1986.

Transition-metal catalysed decomposition of aliphatic diazo compounds, G. Maas, *Top. Curr. Chem.,* 1987, **137**, 75.

Rhodium(II) catalysed reactions of diazocarbonyl compounds, J. Adams and D. M. Spero, *Tetrahedron,* 1991, **47**, 1765.

Ylide formation from the reaction of carbenes and carbenoids with heteroatom lone pairs, A. Padwa and S. F. Hornbuckle, *Chem. Rev.,* 1991, **91**, 263.

4   Nitrenes

Nitrenes are six-electron neutral, monovalent, highly reactive nitrogen intermediates, the nitrogen atom having four non-bonded electrons (two being the 'normal' lone pair associated with nitrogen) indicated by the four dots in the drawn structure. Such monovalent short-lived species were first suggested by Tiemann in 1891 as intermediates in the Lossen rearrangement (see Section 4.5), and were also adopted by Curtius to explain various reactions of azides. Although various names for these intermediates such as imidogen (once favoured by *Chemical Abstracts*), imenes, azenes, have been used in the past, the word nitrene is now in almost universal use, the nomenclature following that of carbenes. Thus PhN is phenylnitrene, $MeSO_2N$ is methanesulphonylnitrene, and so on. The chemistry of nitrenes closely parallels that of carbenes in virtually all aspects, and therefore some points are covered in less detail, since it is assumed that the reader is already familiar with the principles established in the previous chapter.

$R-\ddot{N}{:}$

General nitrene structure

4.1 Structure and reactivity

Like carbenes, there is the possibility of two spin states for nitrenes, depending on whether the two non-bonding electrons (the 'normal' nitrogen lone pair remains paired) have their spins paired or parallel. In general nitrenes obey Hund's Rule and are ground state triplets with two degenerate *sp*-orbitals containing a single electron each, although the nitrogen atom in the singlet is usually represented as sp^2-hybridised. The energy difference between the singlet and triplet states is usually much larger for nitrenes than for carbenes, being estimated at 145 kJmol^{-1} for nitrene (NH) itself compared with 32–42 kJmol^{-1} for methylene (CH$_2$). This is in part due to the fact that nitrogen is more electronegative than carbon, and therefore holds its electrons closer to the nucleus. As expected, the nature of the substituent on nitrogen affects both the multiplicity and the normal electrophilic reactivity of nitrenes. Strong π-donor substituents such as amino groups greatly stabilise the singlet as well as causing the nitrene to exhibit nucleophilic character in its reactions. A dramatic example of this effect is seen in the nitrene derived by oxidation of 1-amino-2,2,5,5-tetramethylpyrrolidine, which is a ground state singlet, stable in solution at low temperature. IR Spectroscopy suggests the presence of a N=N bond ($v_{max.}$ 1638 cm^{-1}), the assignment being confirmed by observing the shift ($v_{max.}$ 1612 cm^{-1}) on isotopic labelling of the exocyclic nitrogen with ^{15}N. Aminonitrenes of this type are somewhat special, and are usually named as 1,1-diazenes, to reflect the fact that they show little nitrene character. Other aminonitrenes, particularly those derived from *N*-aminoheterocycles, however, do show nitrene behaviour.

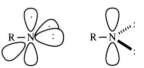

sp triplet and sp^2 singlet nitrenes

Singlet – triplet gap is larger for nitrenes than for carbenes

Stabilisation of singlet nitrene by donor substituents

Although many nitrenes are too unstable to allow the sort of direct observation referred to above, various techniques can be used to obtain the UV and ESR spectra of more transient species. Interestingly, nitrene itself (NH) has been extensively characterised by spectroscopy, the electronic absorption at 336 nm being first observed in 1892, although it was some years before the assignment was unambiguously confirmed. More modern techniques such as laser flash photolysis and low temperature matrix studies allow the routine measurement of UV spectra of nitrenes. Triplet phenylnitrene, for example, shows absorptions between 300 and 400 nm at 77K, whilst ESR measurements on the same species suggest that there is substantial delocalisation of a single electron into the aromatic ring.

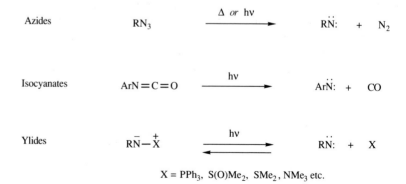

Delocalised structure of triplet phenylnitrene

4.2 Generation of nitrenes

Methods for the generation of nitrenes are summarised below and on the next page, and it will be apparent that, with the exception of the reductive and oxidative routes, the methods closely parallel those used for carbenes.

Azides RN_3 $\xrightarrow{\Delta \; or \; h\nu}$ $R\ddot{N}:$ + N_2

Isocyanates $ArN{=}C{=}O$ $\xrightarrow{h\nu}$ $Ar\ddot{N}:$ + CO

Ylides $R\bar{N}{-}\overset{+}{X}$ $\underset{\longleftarrow}{\xrightarrow{h\nu}}$ $R\ddot{N}:$ + X

X = PPh_3, $S(O)Me_2$, SMe_2, NMe_3 etc.

Methods for the generation of nitrene intermediates

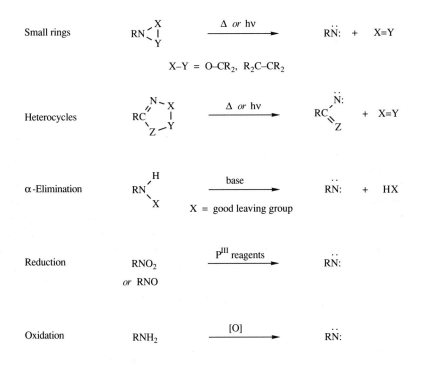

Small rings

$$RN\underset{Y}{\overset{X}{\lessgtr}} \quad \xrightarrow{\Delta \; or \; h\nu} \quad RN: \quad + \quad X{=}Y$$

$$X{-}Y \;=\; O{-}CR_2, \; R_2C{-}CR_2$$

Heterocycles

$$RC\overset{N-X}{\underset{Z-Y}{\lessgtr}} \quad \xrightarrow{\Delta \; or \; h\nu} \quad RC\overset{N:}{\underset{Z}{\lessgtr}} \quad + \quad X{=}Y$$

α-Elimination

$$RN\overset{H}{\underset{X}{\lessgtr}} \quad \xrightarrow[\text{X = good leaving group}]{\text{base}} \quad RN: \quad + \quad HX$$

Reduction

$$\begin{array}{c} RNO_2 \\ \textit{or } RNO \end{array} \quad \xrightarrow{P^{III} \text{ reagents}} \quad RN:$$

Oxidation

$$RNH_2 \quad \xrightarrow{[O]} \quad RN:$$

Methods for the generation of nitrene intermediates (continued)

Azides

Azides, which have been known for over 100 years are the most widely used precursors of nitrenes. Like diazo compounds they possess a linear 1,3-dipolar structure, and are easily prepared, often by introduction of the N_3^- ion from inorganic salts such as sodium azide. The thermal stability of azides is critically dependent on the substituent on nitrogen. In genuine nitrene reactions the loss of nitrogen is rate determining being largely independent of the nature of the solvent and the concentration of any other compounds present. Whereas most azides decompose thermally in the 100–200°C range, some are much less stable, particularly those in which the loss of nitrogen can be assisted in some way. In these assisted reactions, exemplified by *ortho*-substituted aromatic azides, nitrenes are probably not involved at all, e.g. the facile formation of anthranils from *ortho*-azido aromatic ketones, probably proceeds by an electrocyclisation followed by loss of nitrogen.

CAUTION! All azides are potentially explosive and must be handled with care.

$$R-\bar{N}-\overset{+}{N}{\equiv}N$$
$$\updownarrow$$
$$R-N{=}\overset{+}{N}{=}\bar{N}$$
$$\updownarrow$$
$$R-\bar{N}-N{=}\overset{+}{N}$$

1,3-Dipolar nature of azides

Assisted decomposition of an *ortho*-substituted aromatic azide

Cyanonitrene

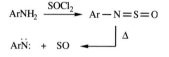

Cyanogen azide (N≡C–N$_3$) is another special case, and is notoriously unstable decomposing at about 50°C. This facile decomposition is probably due to the unique nature of cyanonitrene – a symmetrical highly stabilised triplet species.

Like diazo compounds, the decomposition of azides can be catalysed by protic and Lewis acids and by transition metal salts, although in the latter case, less is known about these processes than the corresponding reactions of diazo compounds. Azides are also readily decomposed photochemically, and this is often the method of choice, particularly for mechanistic studies.

Isocyanates

Aryl isocyanates do not eliminate CO to give nitrenes on heating since the process is energetically unfavourable, although irradiation can provide sufficient energy for the reaction to proceed. Hence the generation of nitrenes from isocyanates is analogous to the generation of carbenes from ketenes. The related sulphinylamines (Ar–N=S=O), readily prepared from anilines and thionyl chloride, do decompose to nitrenes thermally with extrusion of SO.

Ylides

Although the phosphorus and sulphur ylides of nitrogen, called imino-phosphoranes and iminosulphuranes (sulphimides) respectively, are less well known than their carbon counterparts, they do give nitrene products on irradiation. For example, photolysis of the *S,S*-dimethyl sulphimide derived from *N*-phenylbenzamidine gives a high yield of 2-phenylbenzimidazole, presumably via cyclisation of the intermediate imidoylnitrene to the aromatic ring. Thermolysis of the corresponding nitrogen–nitrogen ylide (an aminimide) gives the same product, as does photolysis of 1,5-diphenyltetrazole and 3,4-diphenyl-1,2,4-oxadiazol-5-one, strongly suggestive of a common intermediate, the nitrene. The same product is also formed by oxidation of *N*-phenylbenzamidine using lead(IV) acetate.

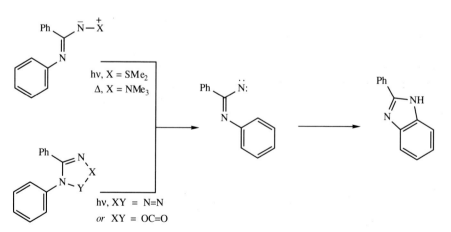

Cyclisation of an imidoylnitrene generated from different precursors

Small rings

Irradiation of oxaziridines results in extrusion of a carbonyl compound and formation of a nitrene in a reaction analogous to the generation of carbenes from epoxides. Oxaziridines can be formed photochemically from nitrones, or by oxidation of imines.

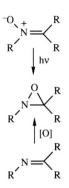

Synthesis of oxaziridines

Photochemical generation of phenylnitrene from an oxaziridine

Heterocycles

Five-membered heterocyclic rings, including those with aromatic stabilisation, that are able to undergo fragmentation with extrusion of nitrogen or carbon dioxide, readily decompose to give nitrenes on irradiation or vapour phase thermolysis. For instance, 1,4,2-dioxazol-5-ones (general scheme, X = Z = O, Y = CO) lose carbon dioxide on heating or irradiation to give acylnitrenes, and examples involving a tetrazole and an oxadiazolone are shown above (p. 54).

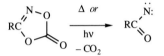

Decomposition of 1,4,2-dioxazol-5-ones

α-Elimination

The historically significant and synthetically useful α-elimination route to carbenes is less important in nitrene chemistry. Although in a few substrates, such as *N,O*-bis(trimethylsilyl)hydroxylamines, the elimination can be effected thermally, more often than not the reaction is base mediated, and requires a good leaving group on nitrogen. Since *N*-halo compounds are often unstable and are prone to radical and ionic reactions, the most useful nitrene precursors of this type are *O*-arenesulphonylhydroxylamines (RNHOSO$_2$Ar), the reaction being best known for ethoxycarbonylnitrene. The elimination can be effected using an organic base such as triethylamine or under phase transfer conditions, and the fact that a genuine nitrene intermediate is involved is strongly suggested by the similar product distribution obtained from α-elimination of EtO$_2$CNHOSO$_2$Ar and thermolysis or photolysis of EtO$_2$CN$_3$.

N,O-bis(trimethylsilyl)hydroxyl-amine

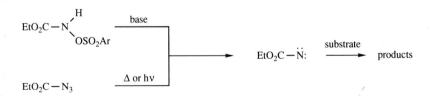

Generation of ethoxycarbonylnitrene by α-elimination; the same products are formed by decomposition of ethyl azidoformate

Reduction of nitro and nitroso compounds

The deoxygenation of nitro and nitroso groups can be carried out with a variety of reagents although trivalent phosphorus compounds, particularly triethyl phosphite, are most commonly used. Although the intermediacy of nitrenes is not proven in every case, the reaction is useful especially for arylnitrenes since the precursor aromatic nitro compounds are so readily accessible. The deoxygenation may involve nucleophilic attack on oxygen by phosphorus followed by loss of the thermodynamically stable P=O compound, although alternatives are possible. The resulting nitroso compound undergoes similar reactions to complete the deoxygenation.

Possible mechanism for phosphorus(III) mediated deoxygenation of nitro and nitroso compounds

Oxidation of amines

The removal of both hydrogens from a NH_2 group by oxidation formally results in generation of a nitrene. Although this is not a practicable route for most nitrenes, the oxidation of primary amines usually leading to many products not necessarily nitrene derived, oxidation of 1,1-disubstituted hydrazines is the method of choice for aminonitrenes. Various oxidants such as MnO_2, HgO, NiO_2 and phenyliodosoacetate [$PhI(OAc)_2$] can be used, but the most commonly employed reagent is lead(IV) acetate. The reactions are carried out at low temperature, typically $-78°C$ in dichloromethane, although in the lead(IV) acetate oxidations there is the strong possibility that the true intermediate is an unstable acetoxy compound, $R_2NNHOAc$, rather than a free nitrene.

$$R_2N-NH_2 \xrightarrow{[O]} R_2N-\ddot{N}:$$

Generation of aminonitrenes by oxidation of 1,1-disubstituted hydrazines

The reaction works particularly well for the 1,1-diazenes, species which exhibit little nitrene character, referred to in Section 4.1, and also for the generation of nitrenes from *N*-aminoheterocycles such as *N*-aminophthalimide. This last reaction is one of the most widely studied oxidative routes to nitrenes, and is useful in that the phthaloyl group can be removed from the subsequent nitrene derived products by reaction with hydrazine hydrate to reveal a free NH_2 group. *N*-Aminoaziridines, prepared by this route are key reagents for the Eschenmoser fragmentation (Section 3.7).

4.3 Cycloaddition reactions of nitrenes

Cycloaddition to alkenes

The formation of aziridines by the 1,2-addition of nitrenes to alkenes mirrors the corresponding reaction of carbenes to give cyclopropanes, although it is

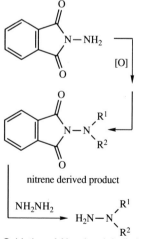

nitrene derived product

Oxidation of *N*-aminophthalimide and removal of the phthaloyl group from the final product with hydrazine

not as general in its scope. Since nitrenes are generally electrophilic the reaction works best for nucleophilic alkenes, and the stereochemistry of the resulting aziridine is usually dependent on the spin state of the nitrene. Thus the Skell hypothesis also holds for nitrenes, although the reaction is subject to the usual experimental variables. For example, thermal decomposition of ethyl azidoformate in neat *cis*-but-2-ene gives the singlet nitrene which adds largely stereospecifically to the alkene to give the *cis*-aziridine. A similar result is obtained when the nitrene is generated by α-elimination from $EtO_2CNHOSO_2Ar$, providing further evidence for the nitrene as a common intermediate in both reactions. However, as the reaction mixture is diluted with an inert solvent, collisional decay to the triplet ground state readily occurs and the reaction becomes less stereospecific. Conversely, addition of triplet (radical) traps such as dienes or α-methylstyrene increases the stereospecificity by selective removal of the triplet. Photolysis is less stereospecific since a higher percentage of the nitrenes are generated directly as triplets.

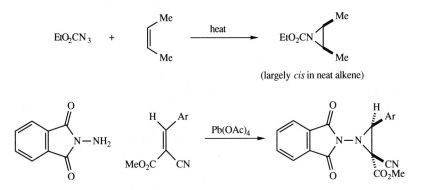

(largely *cis* in neat alkene)

Stereospecific addition of nitrenes to alkenes (in the latter example, an *N*-acetoxy compound rather than a free nitrene is probably involved)

Aminonitrenes such as phthalimidonitrene, being ground state singlets, usually add stereospecifically to alkenes, and because of their somewhat nucleophilic nature also add to electron deficient double bonds. In contrast, arylnitrenes do not usually undergo addition to alkenes, although if aryl azides are used as precursors, a reaction certainly occurs. This turns out to be a general problem in azide/nitrene chemistry since azides are reactive 1,3-dipoles which readily undergo cycloaddition to alkenes to give five-membered heterocycles, 1,2,3-triazolines. These may subsequently lose nitrogen to give aziridines, presumably by way of a diradical intermediate, with the consequent loss of alkene stereochemistry.

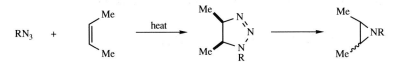

Formation of aziridines by initial 1,3-dipolar cycloaddition of an azide

Cycloaddition to 1,3-dienes

By analogy with carbenes, nitrenes undergo exclusive 1,2-addition to 1,3-dienes to give vinylaziridines, which on heating are transformed into dihydropyrroles. The reaction is known for various nitrenes including ethoxycarbonylnitrene and aminonitrenes, although when azides are used as precursors the non-nitrene 1,3-dipolar azide cycloaddition mechanism may operate. In its intramolecular variant, the reaction has been used as a route to pyrrolizidine alkaloids (see Section 4.7).

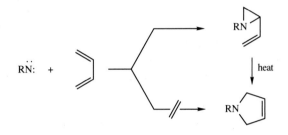

1,2-Addition to 1,3-dienes to give vinylaziridines

Cycloaddition to alkynes

1*H*-azirine   2*H*-azirine

The addition of nitrenes to alkynes immediately presents a problem. The initial product of addition to the triple bond is a 1*H*-azirine, a species which is formally antiaromatic since it is planar and has four delocalisable π-electrons (including the nitrogen lone pair). In the case of phthalimidonitrene the presumed intermediate 1*H*-azirine formed by cycloaddition to 3-hexyne, readily rearranges to the stable 2*H*-isomer.

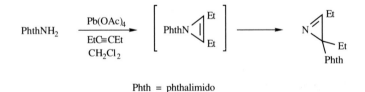

Phth = phthalimido

Addition of phthalimidonitrene to an alkyne to give, after rearrangement, a 2*H*-azirine

Attempts to add nitrenes derived by heating azides to alkynes are usually thwarted by the facile addition of the azide 1,3-dipole to the triple bond with the formation of a stable aromatic 1,2,3-triazole as shown below. Acylnitrenes can also act as 1,3-dipoles, and add to alkynes to give oxazoles in a reaction analogous to the formation of furans from acylcarbenes and alkynes.

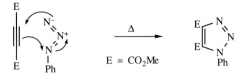

E = CO₂Me

1,3-Dipolar cycloaddition of an azide to an alkyne to give a 1,2,3-triazole

Cycloaddition to arenes

Benzene and its derivatives react with nitrenes to give ring expanded products and/or *N*-substituted anilines. Both types of product are formed by initial nitrene attack on the π-system, presumably by the singlet, to give an azanorcaradiene intermediate, which undergoes electrocyclic rearrangement to the azepine. In the case of ethoxycarbonylnitrene (R = CO₂Et) generated from the azide or by α-elimination, azepines are isolated, although rearrangement to the aniline derivative occurs readily in the presence of acid. The corresponding reaction of sulphonylnitrenes (R = SO₂Ar), even in the absence of acid, usually only gives sulphonyl-anilines which may arise from the azepine or directly from the azanorcaradiene by an alternative pathway due to the enhanced stability of the incipient sulphonamide anion.

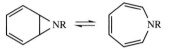

Cope rearrangement

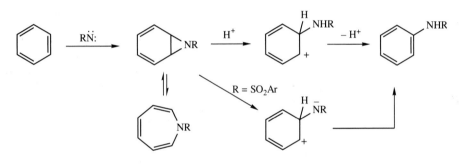

Addition of a nitrene to an aromatic ring

4.4 Insertion reactions of nitrenes

Nitrenes like carbenes readily insert into single bonds. The overall process described by the word insertion can occur by different mechanisms, although most of the preparatively useful C–H insertion reactions are believed to involve direct insertion of the singlet nitrene rather than a stepwise H-abstraction–recombination mechanism involving the triplet. Both mechanisms are shown below. However, many reactions which lead to products with a new C–N bond that are the formal result of C–H insertion proceed by an entirely different mechanism.

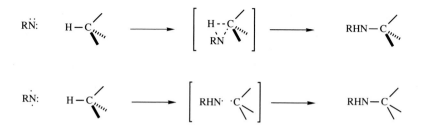

Insertion of singlet and triplet nitrenes into a C–H bond

The importance of the C–H insertion reaction of nitrenes lies in the fact that it is a potentially useful way of functionalising unactivated C–H bonds, converting hydrocarbons into amine derivatives. The reaction is highly dependent on the substituents on nitrogen, cyclohexane, for example, only giving good yields of aminocyclohexane derivatives with acyl-, sulphonyl-, and cyano-nitrenes, generated by thermal or photochemical decomposition of the corresponding azides. Alkylnitrenes give very poor yields of insertion products because of the competing rearrangement by 1,2-hydrogen shift (Section 4.5), and arylnitrenes often give anilines by hydrogen abstraction by the triplet.

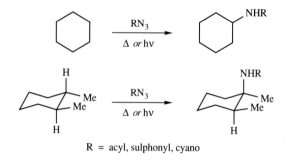

R = acyl, sulphonyl, cyano

Functionalisation of cyclohexanes by nitrene insertion

For all nitrenes studied, the selectivity of C–H insertion follows the expected pattern of reactivity decreasing in the order: tertiary > secondary > primary C–H, although there is considerable variation in the degree of selectivity. For example, the relative reactivities of tertiary, secondary and primary C–H bonds in 2-methylbutane towards ethoxycarbonylnitrene are approximately 30:10:1, whereas the corresponding values for methanesulphonylnitrene are 10:4:1.

The selectivity of nitrene insertion reactions has been widely studied using substituted cyclohexanes as substrates. For arylnitrenes exhibiting triplet reactivity, the reaction is often low yielding and non-stereospecific, but ethoxycarbonyl-, methanesulphonyl-, and cyano-nitrenes all insert with retention of configuration into the tertiary C–H bond of both *cis* and *trans*

1,2-dimethylcyclohexane. When optically active substrates are used, virtually complete retention is again observed, ethoxycarbonylnitrene inserting with 98–100% retention into the tertiary C–H bond of (*S*)-3-methylhexane. The result is independent of the method of nitrene generation, and of concentration, confirming the view that only the singlet species inserts into unactivated C–H bonds.

Although arylnitrenes often give poor yields of intermolecular C–H insertion products, the intramolecular reaction works well, and proceeds with retention of configuration at the reacting centre as shown by the example below (*c.* 100% retention). Intramolecular C–H insertion reactions of other nitrenes have found use in synthesis (Section 4.7).

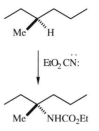

C–H Insertion with retention

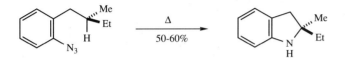

Intramolecular nitrene C–H insertion with retention of configuration

Formal insertions into aromatic C–H bonds often proceed by another mechanism, for example the formation of substituted anilines discussed in Section 4.3. This is also true in intramolecular cases, and the formation of carbazole from biphen-2-ylnitrene, a well studied example, probably occurs by an electrocyclic reaction followed by 1,5-hydrogen shift. The high yielding synthesis of indoles from azidocinnamates proceeds similarly by cyclisation of a vinylnitrene.

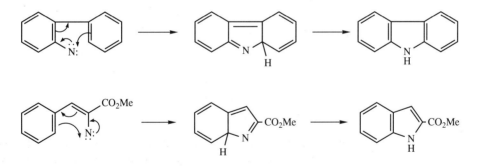

Electrocyclic ring closure of aryl- and vinyl-nitrenes

4.5   Rearrangement of nitrenes

The rearrangement by a 1,2-shift of an atom or group from the adjacent carbon to the electron deficient centre which occurs so readily with carbenes is also a characteristic reaction of nitrenes. When the migrating group is hydrogen, the rearrangement is particularly facile and hence other intermolecular reactions such as cycloaddition involving alkylnitrenes are rarely seen. The rearrangement results in the formation of an imine, which if unsubstituted on nitrogen is easily hydrolysed to the corresponding carbonyl compound. Bearing in mind the fact that azides, which are usually prepared from halides by reaction with inorganic azide, are the most common nitrene sources, the 1,2-rearrangement provides an overall route from halides to carbonyl compounds. Although the mechanism is often depicted as involving a free nitrene (as below), there is rarely evidence to support this, and it likely that as with carbenes, the migration is 'concerted' with the departure of the leaving group from nitrogen in the nitrene precursor.

R = H >> aryl > alkyl

Relative migratory aptitude of groups in nitrene rearrangements

Formation of imines by 1,2-shift of a group to the electron deficient nitrene

The nitrene analogues of the Wolff rearrangement are the well known Curtius, Hofmann, and Lossen reactions. The Curtius rearrangement of acyl azides to isocyanates, and hence amines upon addition of water and decarboxylation, almost certainly does not involve nitrenes, the migration of the group R being concerted with loss of nitrogen in both thermal and photochemical reactions. In the thermal reaction all attempts to trap intermediate nitrenes failed, and although other nitrene derived compounds such as insertion products are formed in the photochemical reaction, these arise from a parallel pathway. The yield of isocyanate is approximately the same whether the nitrene is diverted or not suggesting that there is no pathway from nitrene to isocyanate. Nevertheless despite the mechanistic uncertainty, the reaction is widely used in preparative organic chemistry as a method of converting a carboxylic acid (RCOOH) into a primary amine (RNH_2) or amine derivative ($RNHCO_2R'$) by way of the acyl chloride (RCOCl) and azide ($RCON_3$).

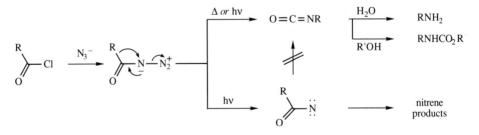

Concerted mechanism for the Curtius rearrangement of acyl azides; the group R migrates with retention of configuration at the migrating carbon

The Hofmann reaction achieves a similar overall transformation, going from an amide (RCONH₂) to an amine (RNH₂) by treatment with bromine under basic conditions. The reaction proceeds by way of an *N*-bromoamide (RCONHBr) which in the presence of base undergoes migration of the R group from carbon to the incipient electron deficient nitrogen concerted with the loss of bromide. The Lossen rearrangement of *O*-acetylhydroxamic acids (RCONHOAc) upon treatment with base no doubt proceeds by a similar process.

Probable mechanism of the Hofmann rearrangement

The rearrangement of arylnitrenes parallels that of arylcarbenes in its complexity, and in the gas phase follows a similar reversible pathway involving a strained seven-membered ketenimine. The reaction is further complicated by the fact that the intermediates may be in equilibrium with 2-pyridylcarbene at high temperatures. The process is extremely complicated, and the final product may be derived from any of the intermediates along the pathway. In solution, the highly reactive intermediates **A** (or **B**) can be intercepted by an added nucleophile such as diethylamine to give, after hydrogen shift, an isolable 3*H*-azepine.

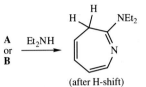

(after H-shift)

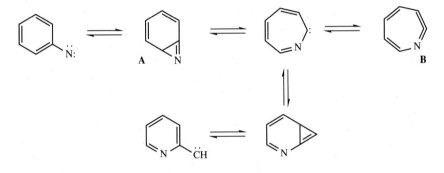

Reversible rearrangement of phenylnitrene showing possible involvement of pyridylcarbene

Finally it should be noted that certain aminonitrenes, because of their unique 1,1-diazene structure, undergo other types of rearrangement and fragmentation reactions. For example, oxidation of 1-alkyl-1-allyl hydrazines leads to azo compounds by [2,3]-sigmatropic rearrangement of the intermediate 1,1-diazene.

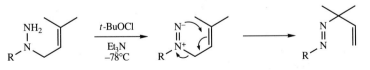

Rearrangement of 1,1-diazene by [2,3]-sigmatropic shift

Sulphoximides, formed by interception of nitrenes by DMSO (Me₂SO)

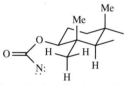

Azo compound; formal nitrene dimer

4.6 Reactions of nitrenes with nucleophiles

Nitrenes for the most part being electron deficient are highly electrophilic intermediates and therefore react with nucleophiles of all types. Tertiary amines, phosphines, sulphides, and sulphoxides all react with nitrenes to give ylides, in a reaction which is the reverse of their formation. In practice, dimethylsulphoxide (DMSO) is often the most convenient nucleophilic trap since it can be used as the reaction solvent, and gives relatively stable sulphoximides.

Azo compounds, which are formally nitrene dimers, are common byproducts in many nitrene reactions. However, the dimerisation of two highly reactive species in solution is extremely unlikely on statistical grounds, and therefore the mechanism of azo compound formation probably involves reaction of a nitrene, as an electrophile, with its precursor.

4.7 Uses of nitrenes in organic synthesis

The major uses of nitrenes in synthesis involve the functionalisation of unactivated C–H bonds by nitrene insertion, and cyclisation reactions to give nitrogen containing rings, particularly those found in alkaloids.

Functionalisation of unactivated C–H bonds

The use of intramolecular nitrene reactions to functionalise unactivated C–H bonds has been widely investigated, particularly in steroids and related systems. For example, in attempts to functionalise the 4,4-dimethyl groups in lanosterol, a process of considerable biosynthetic importance, thermal decomposition of 3β-lanost-8-enyl azidoformate gave a δ-lactone (25%) resulting from insertion into the 4α-methyl group. The major product (55%) was a γ-lactone formed by insertion into the 2α-C–H bond.

Insertion into the 4α-methyl

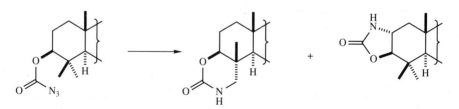

Intramolecular nitrene C–H insertion in functionalisation of a lanosterol derivative

Alkaloid synthesis

The use of nitrene insertion reactions in the synthesis of diterpene alkaloids goes back to the early 1960s. For example, in a reaction similar to that described above, irradiation of the acyl azide shown below, results in intramolecular insertion into the axial methyl group to give a bridged δ-lactam which has the azabicyclononane skeleton of the atisine alkaloids, albeit in poor yield due to competing Curtius rearrangement to give the corresponding isocyanate in about 70% yield.

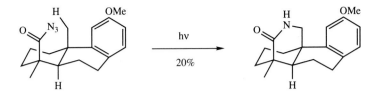

Nitrene route to the atisine alkaloids

The apparent intramolecular insertion of vinylnitrenes into aromatic C–H bonds referred to earlier has found wide use in the synthesis of naturally occurring indoles in one of our (C. J. M.) own laboratories. The precursor azides, easily prepared from benzaldehydes in one step, decompose on heating in boiling toluene or xylene to give indole-2-esters in high yield. The reaction has been used as the key step in the synthesis of natural products such as methoxatin (co-enzyme PQQ) as shown below.

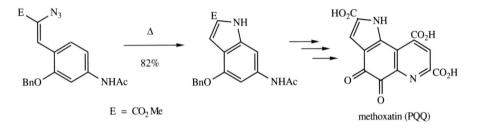

Nitrene cyclisation as the key step in the synthesis of the bacterial coezyme methoxatin

Intramolecular cycloaddition to a 1,3-diene has been used in an elegant approach to pyrrolizidine alkaloids. Heating the dienyl azide results in intramolecular 1,2-addition to give a vinylaziridine; on further heating this undergoes rearrangement to the five-membered ring, and finally simple reduction gives isoretronecanol, the basic ring skeleton of the alkaloids. Of course, a mechanism involving azide cycloaddition rather than a free nitrene cannot be ruled out.

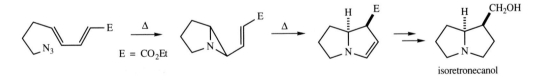

Intramolecular 1,2-addition to a diene in the synthesis of isoretronecanol

Nitrene rearrangements have been used to effect useful ring expansion reactions in which an exocyclic nitrogen, usually in the form of an azide, is subsequently incorporated into the ring by a 1,2-shift. The process is exemplified by the synthesis of nicotine shown below; treatment of the tertiary alcohol with hydrazoic acid gives the corresponding tertiary azide which immediately undergoes loss of nitrogen accompanied by a 1,2-shift to give the imine. Reduction and *N*-methylation then complete the synthesis.

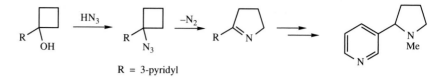

R = 3-pyridyl

Azide mediated ring expansion in a synthesis of nicotine

Problems

1. Write a mechanism for the following transformation and account for the low temperature at which the azide decomposes.

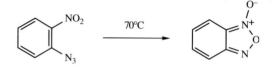

2. Suggest a nitrene based route for the conversion of 1-adamantanol into 2-amino-1-adamantanol. (*Hint:* three steps including one intramolecular reaction are required.)

3. Rationalise the following observation.

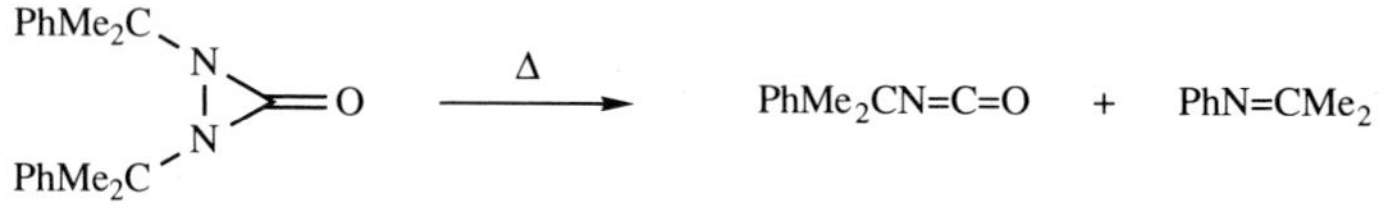

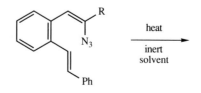

4. Predict the possible product(s) from the following reaction.

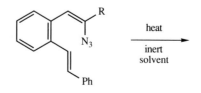

5. 2-Methylcarbazole is formed by all the three routes shown below. Explain these results in terms of the intermediates involved and their subsequent reactions.

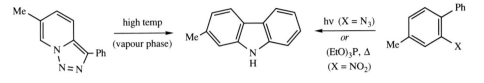

Further reading

Nitrenes, ed. W. Lwowski, Wiley Interscience, New York, 1970.

Reactive molecules: the neutral reactive intermediates in organic chemistry, C. Wentrup, Wiley, New York, 1984.

Azides and nitrenes, ed. E. F. V. Scriven, Academic Press, Orlando, Florida, 1984.

Azides: their preparation and synthetic uses, E. F. V. Scriven and K. Turnbull, *Chem. Rev.,* 1988, **88**, 297.

Vinyl azides in natural product synthesis, C. J. Moody, *Studies in Natural Product Synthesis,* 1988, **1**, 163.

An overview of the total synthesis of pyrrolizidine alkaloids via [4+1]-azide-diene annulation methodology, T. Hudlicky, G. Seoane, J. D. Price and K. G. Gadamselti, *Synlett,* 1990, 433.

Oxidation of unactivated C–H bonds by nitrene insertion, C. J. Moody, *Comprehensive Organic Synthesis,* Pergamon, Oxford, 1991, Vol. 7, p. 21.

5 Arynes

Arynes are neutral intermediates derived from aromatic rings by removing two substituents leaving behind two electrons to be distributed between two orbitals. Although in most cases the substituents are *ortho* to one another, this is not a prerequisite, and *meta* and *para* arynes are possible intermediates. The most common aryne, C_6H_4, is based on benzene itself, although all aromatic and heteroaromatic systems can potentially give similar species. The need to postulate a C_6H_4 intermediate arose in the 1870s in order to explain the formation of biphenyl in certain reactions, and although the first correct formulation of a related intermediate from benzofuran was put forward in 1902, it was not until the 1940s that Wittig advanced clear and convincing arguments for the intermediacy of arynes. This was followed in 1953 by the classic isotopic labelling experiments of Roberts (see Section 5.2) and the Diels–Alder trapping by Wittig in 1955 (see Section 5.4) which finally confirmed the existence of benzyne.

The *aryne* nomenclature derives from the fact that the *ortho*-isomer of C_6H_4 (*ortho*-benzyne) can be represented as an alkyne (see below), although systematically the species should be named as a didehydroaromatic compound. Hence *ortho*-benzyne is 1,2-didehydrobenzene and 3,4-pyridyne is 3,4-didehydropyridine and so on. Nevertheless the aryne nomenclature remains in common use and will be used here throughout. Although many arynes have been implicated as intermediates, by far the most studied species is *ortho*-benzyne, a fact reflected by the remainder of this chapter.

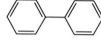

Biphenyl

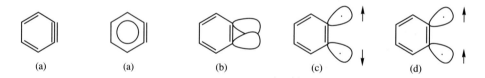

(a)　　(a)　　(b)　　(c)　　(d)

Representations of *ortho*-benzyne showing (a) the alkyne structure resulting from (b) lateral overlap of the two *p*-orbitals, and (c) singlet and (d) triplet diradical structures

5.1 Structure and Reactivity

ortho-Benzyne

ortho-Benzyne is usually represented as a singlet molecule with a carbon–carbon triple bond. This strained π-bond is formed by lateral overlap of the two orbitals in the plane of the ring. The alternative structure is a diradical, either singlet or triplet, although most of the chemistry of the species is in accord with the alkyne structure.　Theoretical calculations support the view that the symmetrical triple bond structure is lowest in energy, and suggest that the structure is distorted from benzene with estimates for C≡C bond length in the range 1.25–1.34Å (cf. 1.20 for ethyne and 1.39 for benzene). The geometric distortion is compensated for by the greater lateral *p*-orbital overlap, although this overlap is still relatively poor compared with that in a normal π-bond. A direct consequence of the strained 'alkyne' bond is that arynes have low lying LUMOs and hence the energy gap between the HOMO and LUMO is small. In accord with this, *ortho*-benzyne shows the properties of a highly reactive alkyne participating in a range of cycloaddition reactions. In other reactions, however, *ortho*-benzyne resembles a carbene, having the similar electronic arrangement of two electrons distributed between two orbitals, and behaves as a powerful electrophile, a second consequence of its low lying LUMO. Recently with the advent of sophisticated experimental techniques, benzyne has been observed directly using laser flash photolysis to obtain its UV spectrum and matrix isolation to record the IR spectrum. In the latter experiment an intermediate, controversially assigned as *ortho*-benzyne, with an IR absorption of 2085 cm^{-1}, typical of a C≡C bond, was observed.

The triple bond in *ortho*-benzyne can be stabilised by complexation to transition metals. Aryne–metal complexes were originally proposed as intermediates in the decomposition of various aryl derivatives of early transition metals, and the first fully characterised mononuclear *ortho*-benzyne complex, TaMe$_2$(η^5-C$_5$Me$_5$)(η^2-C$_6$H$_4$), was prepared thus. Although this method does not appear general for all transition metals, various complexes of zirconium, rhenium and niobium have been characterised. More recently complexes of nickel and platinum have been prepared. For example, oxidative addition of 1,2-dibromobenzene to a nickel(0) complex of ethene followed by reduction with sodium amalgam gives the *ortho*-benzyne nickel complex. The same complex can be prepared by generating *ortho*-benzyne from fluorobenzene in the presence of the nickel ethene complex. X-Ray crystallographic analysis shows that the metal atom and the two aryne carbons form a metallocyclopropene ring.

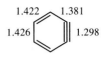

One of the calculated structures for *ortho*-benzyne (bond lengths in Å)

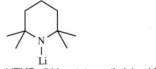

LiTMP - lithium tetramethylpiperidide

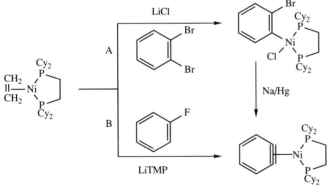

Preparation of nickel complex of *ortho*-benzyne (Cy = cyclohexyl); route A involves oxidative addition of 1,2-dibromobenzene to the nickel complex, followed by reduction, whereas route B involves displacement of ethene from the nickel complex by the free aryne.

More impressive still, perhaps, is the successful preparation of benzdiyne complexes. Benzdiyne is unknown in the free state but can be stabilised on a transition metal, and zirconium, nickel and platinum complexes have been fully characterised. Although aryne transition metal complexes undergo a range of fascinating reactions, their potential in organic synthesis remains to be fully exploited.

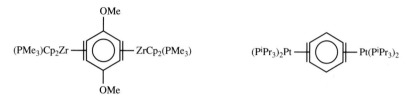

Zirconium and platinum complexes of benzdiyne (Cp = η^5-C$_5$H$_5$)

meta-Benzyne and *para*-benzyne

The removal of two substituents from the *meta*- and *para*-positions of a benzene ring leaving two electrons in two orbitals should generate *meta*- and *para*-benzyne respectively. The benzyne nomenclature is still used for these intermediates despite the fact that there is no formal C≡C bond present, and the species are more correctly named as 1,3- and 1,4-didehydrobenzene. The structures can be represented either as diradicals or bicyclic compounds with 1,3- and 1,4-bonding. In the case of *meta*-benzyne, calculations suggest that the bicyclic species, bicyclo[3.1.0]hexa-1,3,5-triene, although highly strained is more stable than the 1,3-diradical, and some experimental evidence has been obtained (Section 5.2).

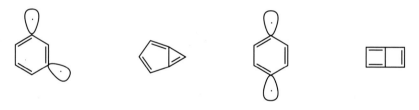

Structures of *meta*- and *para*-benzyne

In contrast, the diradical form of *para*-benzyne appears to be lower in energy than the highly strained bicyclo[2.2.0]hexa-1,3,5-triene (butalene). Evidence for the existence of the diradical, often termed benz-1,4-diyl, was obtained by Bergman in the early 1970s in the specific deuterium scrambling that was observed on heating 1,6-dideuteriohexa-1,5-diyn-3-ene. The intermediate 1,4-diradical also abstracts chlorine from CCl$_4$ to give largely 1,4-dichlorobenzene.

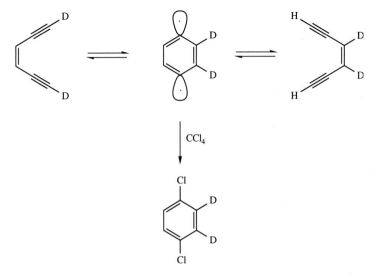

The Bergman cyclisation of enediynes to give *para*-benzyne (benz-1,4-diyl)

Although the so-called Bergman cyclisation has been known for over twenty years, it has very recently attained renewed significance, since a similar cyclisation is believed to be the key step in the biological mechanism of action of the calicheamicin/esperamicin and related enediyne antitumour agents. Upon biological activation, the enediyne cyclises to give an intermediate *para*-benzyne which is thought to abstract hydrogen from the DNA sugar backbone and hence prevent replication.

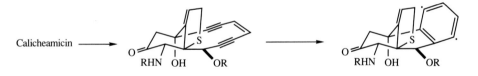

Involvement of a *para*-benzyne in the mechanism of action of the anticancer antibiotic calicheamicin

5.2 Generation of arynes

The main methods of aryne generation are summarised below, and although the examples are limited to *ortho*-benzyne, the relevance of the particular route to other arynes will be referred to in the text.

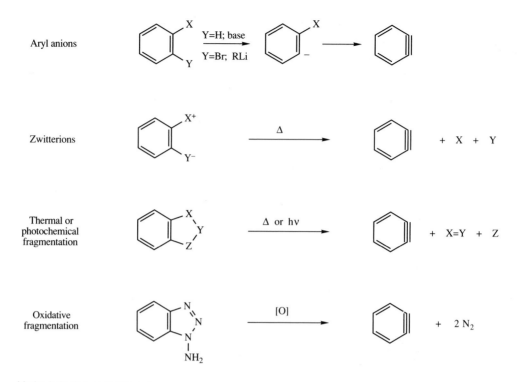

Methods for the generation of arynes

Aryl anions

The elimination of a leaving group from the *ortho*-position of a metallated aromatic ring is historically the most important route to arynes. Although arynes had long been suggested as intermediates in such reactions, it was not until the 1940s that Wittig put the subject on a clear experimental footing. In studying the formation of biphenyl from the reaction of halobenzenes with phenyllithium, he found that the rate was fastest for fluorobenzene. Since this was not the expected result for a simple displacement reaction (fluoride is a worse leaving group than chloride or bromide), an alternative mechanism was suggested whereby the strongly electronegative fluorine facilitated removal of the *ortho*-proton to give 2-fluorophenyllithium. (In modern terminology we would describe this as a directed lithiation reaction.) Loss of lithium fluoride from this formal anion generates benzyne which rapidly reacts with the strongly nucleophilic phenyllithium. Finally quenching with water in work-up gives biphenyl.

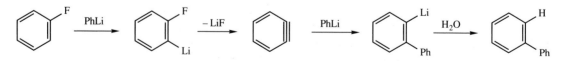

Formation of biphenyl from PhF and PhLi showing the involvement of *ortho*-benzyne

The ^{14}C-labelling experiments of Roberts in 1953 put the existence of *ortho*-benzyne as an intermediate beyond doubt when it was found that treatment of 1-^{14}C-chlorobenzene with potassamide in liquid ammonia gave a 1:1 mixture of 1- and 2-^{14}C labelled aniline. The reaction had clearly proceeded through a symmetrical intermediate, *ortho*-benzyne. The overall process whereby a nucleophile apparently enters *ortho* to the leaving group is referred to as *cine substitution*.

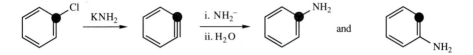

The Roberts isotopic labelling experiment: the shaded atom represents [14]C

The *ortho*-metallation of aromatic halides followed by loss of metal halide is now a firmly established route to *ortho*-arynes. Several bases can be used: alkali metal amides (LiNH$_2$, NaNH$_2$ etc.), alkyllithiums, lithium amides (LiNR$_2$), potassium *tert*-butoxide etc. Some examples of *ortho*-benzyne generation are shown below. The use of trifluoromethanesulphonate (triflate, OTf) as a superior leaving group to halide is a noteworthy modern advance, particularly when combined with the use of the trimethylsilyl group which on treatment with fluoride ion gives the aryl anion. Precursors which can give aryl anions by other mechanisms are also potential sources of arynes, for example, base induced cleavage of aryl ketones.

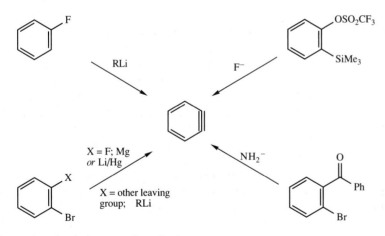

Generation of *ortho*-benzyne via aryl anions

In cases where two possible arynes could be formed from a single precursor, the result will depend on the relative rates of the two steps involved in aryne generation. If the formation of the anion is rate determining then this will control which aryne is formed. For example, lithiation of 3-methoxybromobenzene occurs at the 2-position since it is directed by the oxygen substituent, and provided that loss of bromide is fast, a single benzyne will result. Likewise powerful *ortho*-directing substituents such as the oxazoline group can be used to control aryne formation.

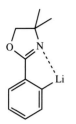

In oxazolines, the lithiation is 'directed' by a combination of the electron withdrawing substituent and chelation to N

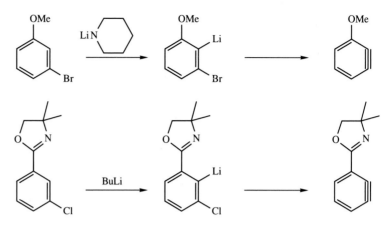

Formation of unsymmetrical arynes

The elimination of the elements of HX by strong base can be extended to *meta*- and *para*-benzynes. Thus there is evidence for the involvement of bicyclo[3.1.0]hexa-1,3,5-triene (*meta*-benzyne) on treatment of the appropriate bicyclic dibromide with potassium *tert*-butoxide. By a combination of isotopic labelling experiments it was shown that only the bicyclic form of *meta*-benzyne, rather than the diradical, was an intermediate in the reaction. The aryne is trapped by nucleophiles such as dimethylamine to give an intermediate which undergoes ring opening to give the fulvene as the final product.

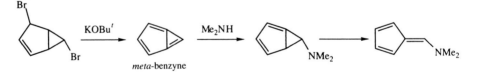

Generation and nucleophilic addition to *meta*-benzyne (bicyclo[3.1.0]hexa-1,3,5-triene)

As we have already seen, the diradical form of *para*-benzyne can be approached by thermal cyclisation reactions of enediynes. The chemical approach to the bicyclic form (butalene) involves a base mediated elimination reaction. Reaction of chlorobicyclo[2.2.0]hexadiene with $LiNMe_2$ at $-10°C$ generates butalene which can be trapped by dimethylamine to give an unstable compound which ring opens to give dimethylaniline. Deuterium labelling experiments, however, show that the reaction is somewhat more complicated than the above analysis would suggest, and for example, direct displacement of chloride cannot be ruled out.

Generation and nucleophilic addition to *para*-benzyne (bicyclo[2.2.0]hexa-1,3,5-triene)

Zwitterions

Probably the most important of all precursors to *ortho*-benzyne is benzenediazonium-2-carboxylate. The compound is easily prepared by diazotisation of 2-aminobenzoic (anthranilic) acid using nitrous acid, and can be isolated as a solid provided it is kept moist. On no account should one attempt to obtain the salt as a dry solid because of the potential explosion risk. To avoid this problem it is quite common to carry out an *in situ* aprotic diazotisation of anthranilic acid using pentyl nitrite ($C_5H_{11}ONO$). When heated in a solvent to about 80°C, nitrogen and carbon dioxide are evolved and *ortho*-benzyne is formed. The decomposition can also be carried out photochemically, and flash photolysis of the diazonium carboxylate has been used to obtain the UV spectrum of *ortho*-benzyne. Benzenediazonium-3- and 4-carboxylates have been studied under similar conditions and UV spectra have been obtained for the transient intermediates so generated.

Benzenediazonium-2-carboxylate
CAUTION – explosion risk

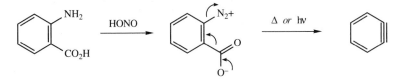

Formation and decomposition of benzenediazonium-2-carboxylate to give *ortho*-benzyne

Although the decomposition of benzenediazonium-2-carboxylate can be conveniently represented as a concerted process (as above), this is almost certainly not the case since there is evidence to suggest that nitrogen is lost first. The diazonium carboxylate method is quite general; various substituted *ortho*-benzynes, 3,4-pyridyne and 2,3-naphthalyne can all be generated from the appropriate amino acid, although some of these may be difficult to prepare, by diazotisation and thermal decomposition. In cases where benzenediazonium carboxylates prove too thermally unstable, one can use a masked version of the diazonium group in the form of a 1,2,3-triazene. These compounds, which should be handled with care since they are potential carcinogens, decompose smoothly at about 130°C (or lower in the presence of acid) with loss of dimethylamine, nitrogen and carbon dioxide.

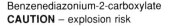

A masked form of
benzenediazonium-2-carboxylate
CAUTION – possible carcinogen

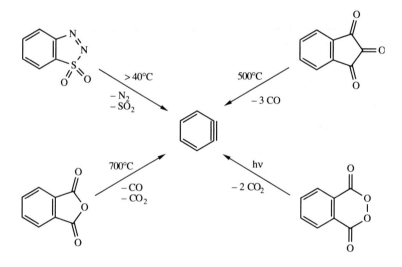

Diphenyliodonium-2-carboxylate

The only other useful aryne precursor of the zwitterionic type is diphenyliodonium-2-carboxylate, prepared from 2-iodobenzoic acid. It is much more thermally stable than benzenediazonium-2-carboxylate decomposing in the range 160–200°C with elimination of carbon dioxide and iodobenzene.

Thermal and photochemical fragmentation of cyclic systems

The formation of *ortho*-benzyne by fragmentation of cyclic systems by electronic rearrangement will only be energetically favourable if the other fragments so formed are extremely thermodynamically stable molecules such as CO, CO_2 and N_2. Even so the thermal energy required to initiate the fragmentation is often considerable, high temperature gas phase reactions being necessary. A clue as to whether such a fragmentation is feasible can often be obtained from the mass spectrum of the precursor – if the molecule does not fragment in the desired fashion under electron impact it is unlikely to do so under 'laboratory conditions.' Some of the successful cyclic precursors to *ortho*-benzyne are shown below, and illustrate the large differences in temperatures required to effect the ring fragmentation according to the stability of the precursor. None of the routes are preparatively useful.

Formation of *ortho*-benzyne by ring fragmentation reactions

Oxidative fragmentation of 1-aminobenzotriazoles

Oxidation of 1-aminobenzotriazole with lead(IV) acetate, nickel peroxide or phenyliodosoacetate at −78°C in dichloromethane results in evolution of nitrogen and the generation of *ortho*-benzyne. The reaction, originally studied by C. W. Rees and co-workers, constitutes an extremely mild route to *ortho*-benzyne, and has been extended to other arynes, notably 1,8-didehydro-naphthalene, a 1,3-diradical. The mechanism formally involves generation of an aminonitrene followed by fragmentation with loss of two molecules of nitrogen, although in the light of recent results, the intermediate in the lead(IV) acetate reaction should probably be formulated as an *N*-acetoxy compound.

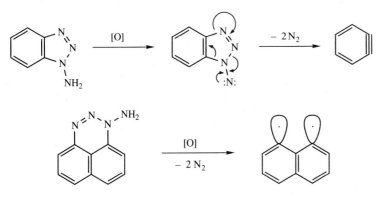

Oxidative fragmentation of *N*-aminoheterocycles

Other methods

There are many other reactions which possibly involve aryne intermediates. Whilst some are mechanistic curiosities, some have been studied in detail, and although none are generally synthetically useful, some examples are included here. Irradiation of 1,2-di-iodobenzene (or 2-iodophenylmercury(II) iodide) can lead to *ortho*-benzyne derived products, probably via an aryl radical intermediate resulting from cleavage of the weak C–I (or C–Hg) bond. Aryl cations, formed by the decomposition of diazonium salts, are also possible intermediates *en route* to *ortho*-benzynes; provided that a large *ortho*-substituent is present, loss of a proton to give an aryne becomes competitive with the normal nucleophilic addition to the cation.

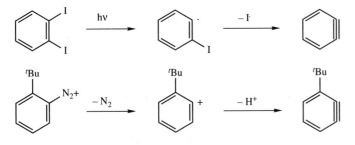

Formation of *ortho*-benzynes via aryl radical and aryl cation intermediates

5.3 Nucleophilic addition to arynes

The formulation of arynes as compounds with a reactive triple bond often disguises the fact that, like carbenes, they are electron deficient species. They are highly electrophilic and therefore react with nucleophiles of all types, although because of their low lying LUMOs and highly polarisable orbitals arynes are 'soft' electrophiles and therefore usually react preferentially with soft nucleophiles. Thus it can be shown that the relative reactivity of various nucleophiles follows the order: $R_3C^- \approx RS^- > $ enolates $> RO^- > I^- > Br^- > Cl^-$. This facile nucleophilic addition has already been seen in the formation of biphenyl when *ortho*-benzyne is generated by the reaction of fluorobenzene with phenyllithium. Indeed addition of aryllithiums to *ortho*-benzynes is a good way of making biaryls. Alkyllithiums add similarly as do acetylenic and benzylic carbanions. Enolates and anions stabilised by cyano, nitro and sulphinyl groups also add to *ortho*-benzyne and some examples are shown below. Many of these anions are used widely in synthesis, and their arylation by arynes, usually carried out by treating a mixture of the aryl halide and the conjugate base of the anion with potassamide in liquid ammonia, is a useful reaction.

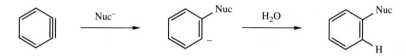

Addition of carbon nucleophiles to *ortho*-benzyne: for PhLi (70%); PhC≡C⁻K⁺ (26%); Ph₃C⁻K⁺ (46%); (EtO₂C)₂CH⁻Na⁺ (51%); ᵗBuO₂CCH₂⁻Na⁺ (40%)

Other nucleophiles of the type RXH (water, alcohols, carboxylic acids, thiols, primary and secondary amines) add readily to *ortho*-benzyne to give the corresponding phenylated compounds, RXPh, often in good yield. In neutral conditions where the nucleophile RXH is not deprotonated, a two step mechanism involving nucleophilic attack followed by proton transfer probably operates.

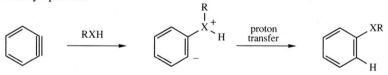

Addition of heteroatom nucleophiles to *ortho*-benzyne: e.g. RXH = H₂O, ROH, RCO₂H, RSH, RNH₂, R₂NH

ortho-Benzyne also undergoes attack from tertiary amines, phosphines, sulphides and related nucleophiles. However in these cases the first formed intermediate cannot undergo the simple proton transfer shown above and therefore reacts in other ways. For example if one of the groups attached to the heteroatom carries a β-hydrogen atom then elimination through a cyclic transition state is the usual process. If no β-hydrogen is available but there are still α-hydrogens present, rearrangement to an ylide is the most likely pathway. Examples of both processes are shown below.

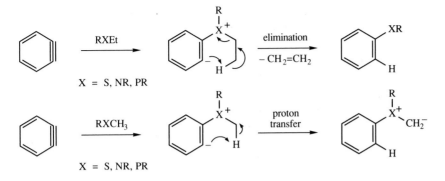

Addition of tertiary amines, phosphines and sulphides to *ortho*-benzyne showing (top) β-elimination and (bottom) ylide formation

All the examples discussed to date have involved *ortho*-benzyne itself, and have simply served to illustrate the range of nucleophiles that react. However one needs to know what happens when unsymmetrical arynes are attacked by nucleophiles, where there are two possible products. The problem arises with substituted *ortho*-benzynes, arynes with an additional ring, or hetarynes. The ratio of products formed depends on the electronic, and to a lesser extent, steric effects of the substituents (or additional ring or heteroatom) and the incoming nucleophile. Since the relevant orbitals of the aryne are orthogonal to the π-system, the electronic effect of substituents is mainly inductive, i.e. is only relayed through the σ-bonds. If the inductive effect of the substituent R can stabilise one of the two transition states to a greater extent then regioselective addition of a nucleophile may be possible. For example, in a 3-substituted aryne, if the substituent is inductively electron withdrawing (−I), such as methoxy or dimethylamino, it will stabilise the developing negative charge *ortho* to itself and hence nucleophilic attack at the *meta*-position is favoured. Conversely if a +I substituent such as methyl is present, nucleophilic attack often occurs at the *ortho*-position. An alternative view is that the reaction is under thermodynamic control and simply proceeds to give the more stable aryl anion intermediate. Strictly of course neither analysis of substituent effects is correct, and one should consider the frontier molecular orbitals of the aryne, but such a rigourous treatment is beyond the scope of this book.

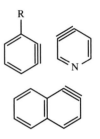

Unsymmetrical arynes; 3,4-didehydropyridine is described as a *hetaryne* since it is derived from a heteroaromatic system

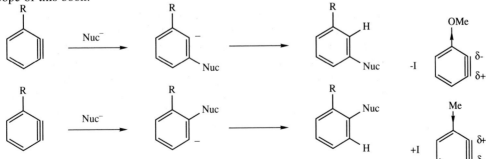

Addition of a nucleophile (Nuc⁻) to an aryne with (top) −I and (bottom) +I substituents

A substituent in the 4-position of an *ortho*-benzyne has a significantly smaller effect presumably because it is further away from the reacting centre. The results of the addition of lithium piperidide to various unsymmetrical arynes are shown below.

Lithium piperidide

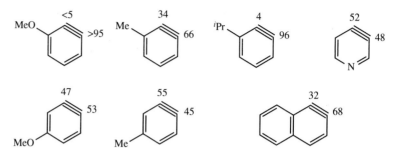

Selectivity in the addition of lithium piperidide to various unsymmetrical arynes: the numbers indicate the relative % addition to each position

Steric effects only become important when either the attacking nucleophile or the substituent *ortho* to the aryne bond is very bulky. An example of this effect is seen above in the increased regioselectivity of addition of lithium piperidide to 3-isopropylbenzyne over 3-methylbenzyne. Likewise in the addition to 3-methylbenzyne, KNH$_2$ adds to both *ortho*- and *meta*-positions in approximately equal amounts whereas the bulkier nucleophile KNPh$_2$ adds exclusively to the less hindered *meta*-position. In additions to 2,3-naphthalyne the *peri*-hydrogen (H-8) hinders the 1-position, and as the bulk of the nucleophile increases so does the selectivity for attack at C-2.

One way to avoid the problem of competing sites of nucleophilic addition is to make the reaction intramolecular. In fact intramolecular nucleophilic addition to arynes has been developed into a useful synthetic method for the preparation of benzo fused ring systems. The substrate is usually an aryl halide bearing a side chain, which contains the potentially nucleophilic centre, in the *ortho*- or *meta*-position. On treatment with base the aryne is generated and simultaneously the side chain is deprotonated, and provided the anion so formed is nucleophilic enough to compete with the external base, cyclisation occurs in good yield. Some examples are given below. Note that four-, five- and six-membered carbocyclic and heterocyclic rings are readily formed: in the last example the alkyl side chain is sufficiently long to allow cyclisation to occur at both ends of the aryne although the *meta*-product is the major one. Benzene rings bridged across their *meta*- and *para*-position are known as *meta*- and *para*-cyclophanes respectively.

2,3-Naphthalyne indicating the *peri*-hydrogen

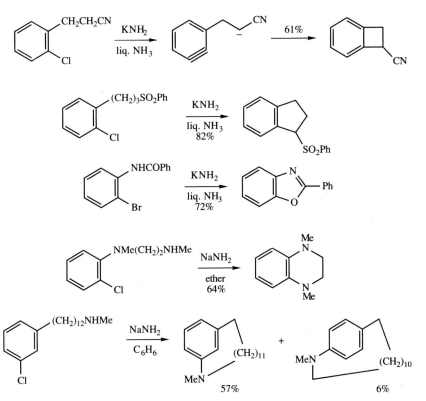

Formation of benzo fused ring systems by intramolecular nucleophilic attack on arynes

5.4 Cycloaddition reactions of arynes

Arynes with their reactive triple bond would be expected to participate readily in cycloaddition reactions. However as demonstrated in the previous section, the addition of nucleophiles is extremely facile, and therefore reactions with non-nucleophilic reagents cannot usually be observed unless the aryne is generated in the absence of nucleophiles. In practice this usually means that routes involving the treatment of aryl halides with nucleophilic bases cannot be used. The first cycloaddition reaction of *ortho*-benzyne, the Diels–Alder reaction with furan, was observed in 1955 by Wittig and used 2-fluorobromobenzene as the precursor. The cycloadduct was obtained in almost 90% yield, and the reaction has formed the basis for a large number of synthetically useful Diels–Alder cycloadditions involving arynes.

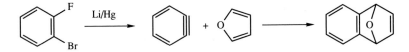

Diels–Alder reaction of *ortho*-benzyne with furan, first observed by Wittig in 1955

Cycloaddition to 1,3-dienes; Diels–Alder reactions

ortho-Benzyne is an extremely reactive dienophile and reacts with a large range of 1,3-dienes to give Diels–Alder products. The reaction is often used as a test for the presence of an aryne intermediate. Furan is commonly used as diene since it is compatible with a number of organometallic reagents and, because it is volatile and easily removed, can be used in large excess. If the aryne is generated under high temperature conditions it is usually better to use the stable but reactive diene 1,2,3,4-tetraphenylcyclopenta-1,3-dien-5-one (tetracyclone). In either case the isolation of the appropriate Diels–Alder adduct, which in the latter case aromatises by loss of CO, is taken as evidence for the intermediacy of an aryne, although it should be noted that such trapping experiments are not always completely unambiguous. The concerted nature of the Diels–Alder reaction is shown by the stereospecific addition of *E,E*-hexa-2,4-diene. In addition to its reactions with 'standard' 1,3-dienes, *ortho*-benzyne undergoes Diels–Alder reaction with compounds not normally considered as dienes such as benzene. Halogenated arynes such as tetrachloro- and tetrafluoro-benzyne are more reactive dienophiles still, and react readily with thiophene and substituted benzenes. Many of these reactions lead to otherwise inaccessible bridged ring systems and are therefore useful in synthesis. Further examples of the Diels–Alder reactions of arynes are given in Section 5.6.

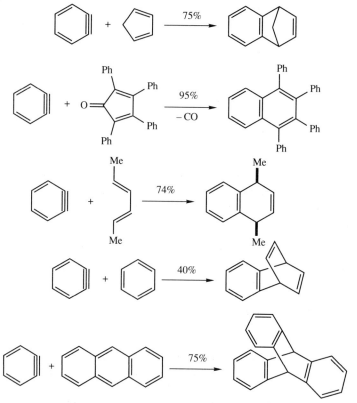

Diels–Alder reactions involving *ortho*-benzyne

Cycloaddition to alkenes

Arynes react readily with simple alkenes to give either benzocyclobutenes or substituted benzenes. The formation of benzocyclobutenes by what is formally a [2+2]-cycloaddition reaction of the aryne to the alkene proceeds best for strained and electron rich C=C double bonds. For example, dicyclopentadiene reacts to give the *exo*-isomer of the corresponding four-membered ring in good yield. The addition to cyanoethene (acrylonitrile) and the reaction with the electron rich ethoxyethene (ethyl vinyl ether) gives the cyano- and ethoxy-benzocyclobutenes in 20 and 40% yield respectively, although the latter reaction almost certainly involves nucleophilic addition of the enol ether to the electrophilic aryne followed by collapse of the betaine intermediate. Likewise, addition to *E*- or *Z*-1,2-dichloroethene is non-stereospecific in accord with a stepwise, probably diradical, mechanism.

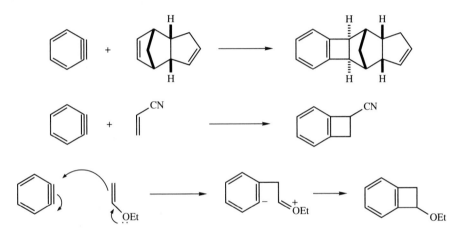

Formation of benzocyclobutenes by addition of *ortho*-benzyne to alkenes

As a reactive dienophile *ortho*-benzyne also participates in the ene reaction. Thus alkenes with an allylic hydrogen can undergo concerted reaction to give substituted benzenes. However the yields are rarely good, the example below proceeding in 17%.

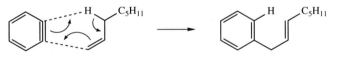

Ene reaction of *ortho*-benzyne

Cycloaddition of *ortho*-benzyne to alkynes should in principle give benzocyclobutadienes. Such intermediates are highly unstable and not surprisingly are not isolated. Instead, the products, formed in low yield, derive from further reaction with another molecule of *ortho*-benzyne or by dimerisation.

1,3-Dipolar cycloaddition

The high reactivity of *ortho*-benzyne is also evident in 1,3-dipolar cycloadditions. The reaction is an extremely useful route to benzo-fused five-membered ring heterocycles, and some examples are given below. Thus azides give benzotriazoles, diazo compounds give, after hydrogen migration, indazoles and nitrile oxides give benzisoxazoles all in good yield.

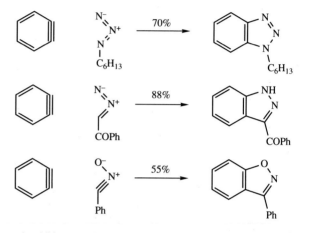

1,3-Dipolar cycloaddition reactions of *ortho*-benzyne (for convenience the 1,3-dipoles are shown as bent; they are, of course, all linear)

5.5 Other reactions

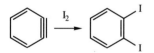

Reaction of *ortho*-benzyne with iodine

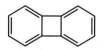

Biphenylene

Although the vast majority of stepwise polar additions to *ortho*-benzyne involve nucleophilic attack on the aryne, electrophilic attack is also possible provided that the aryne is generated by a method which does not involve strongly basic conditions. Few such additions are synthetically useful, with the exception of the formation of 1,2-dihalobenzenes by reactions of *ortho*-benzynes with halogens, although alternative mechanisms initiated by nucleophilic attack of halide may be envisaged. Radical reactions of *ortho*-benzyne, on the other hand, are extremely rare, since as we have already seen in Section 5.1, virtually all of the chemistry of *ortho*-benzyne is associated with the 'triple' bond rather than the diradical form.

   Many reactions involving *ortho*-benzyne produce its dimer and trimer, biphenylene and triphenylene respectively. Although in some cases these hydrocarbons are formed as by-products, in others the yields can be quite high. For example, if *ortho*-benzyne is generated in solution by oxidation of 1-aminobenzotriazole in the absence of a suitable trap, biphenylene is formed in 85% yield. The reaction is highly efficient presumably because the aryne is generated in high local concentration, and therefore dimerises before it can react by other pathways. Decomposition of benzenediazonium-2-carboxylate can also give respectable amounts of biphenylene, although both these precursors only give traces of triphenylene. High yields (over 80%) of triphenylene can be obtained however if *ortho*-benzyne is generated from an aryl halide. It is extremely unlikely that under these conditions in the presence of nucleophiles a concerted trimerisation occurs. Much more likely is a stepwise mechanism which involves reaction of *ortho*-benzyne with the nucleophilic *ortho*-metallated aryl halide to give an intermediate metallated biphenyl which can react with a second molecule of *ortho*-benzyne as shown to give, eventually, the observed trimer.

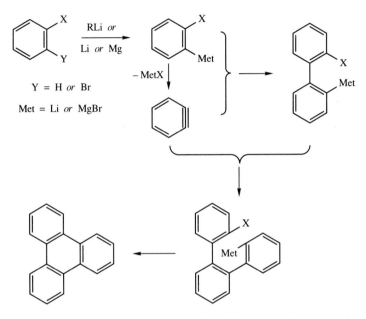

Formation of triphenylene by stepwise nucleophilic additions to *ortho*-benzyne

5.6 Uses of arynes in organic synthesis

Arynes have not been exploited in organic synthesis to the same extent as radicals or carbenes, but nonetheless do undergo some useful transformations. At the simplest level arynes are intermediates in the preparation of 'unusual' isomers of substituted aromatic compounds. For instance treatment of 2-methoxybromobenzene with potassamide in liquid ammonia will result in the formation of the aryne, which will undergo regioselective nucleophilic attack *meta* to the methoxy group to give 3-methoxyaniline. This is an 'unusual' isomer in the sense that conventional routes such as nitration of anisole followed by reduction of the nitro group do not give the *meta*-isomer.

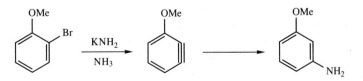

Cine-substitution reaction involving an aryne intermediate

See Section 5.3 for an explanation of the regioselectivity of nucleophilic addition

The major application of arynes in synthesis is in the construction of polycyclic systems using either the Diels–Alder or intramolecular nucleophilic addition reactions. As we have already seen in Section 5.4, the scope of the Diels–Alder reactions with arynes is extremely wide and compounds such as aryl-alkenes and -alkynes, not normally reactive 1,3-dienes, readily participate. Thus 3,4-dimethoxyphenylethyne reacts with *ortho*-benzyne to give, after hydrogen migration, the dimethoxyphenathrene in 42% yield. A similar reaction has been used in an approach to the aporphine alkaloids, whereby 3-methoxybenzyne undergoes Diels–Alder reaction with an aryl alkene to give, again after hydrogen migration, the tetracyclic ring system in 30% yield.

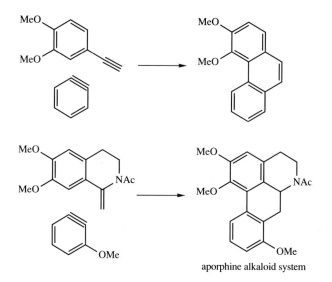

aporphine alkaloid system

Diels–Alder reactions of *ortho*-benzynes in the synthesis of polycyclic systems

An intermolecular Diels–Alder reaction of 3,4-pyridyne has been used in a short synthesis of the important anticancer alkaloid ellipticine. In this case the diene is an α-pyrone; the initial Diels–Alder adduct is not isolated since it spontaneously aromatises by loss of carbon dioxide. Unfortunately the Diels–Alder reaction is not regioselective and an equal amount of the product arising from the alternative direction of addition to 3,4-pyridyne is formed.

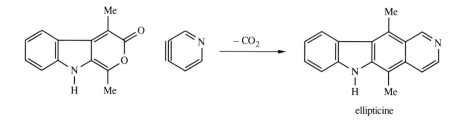

ellipticine

Synthesis of the antitumour alkaloid ellipticine by Diels–Alder addition to 3,4-pyridyne

Two further examples of the use of arynes in alkaloid synthesis are given below, both involving intramolecular nucleophilic addition reactions. In the first case, treatment of the phenolic bromide with the sodium salt of dimethyl sulphoxide as base results in the simultaneous formation of the phenoxide anion and the aryne in the lower ring. Intramolecular nucleophilic attack then gives the alkaloid cularine. The yield is modest (20%) due to competing attack on the aryne by the nucleophilic nitrogen of the isoquinoline.

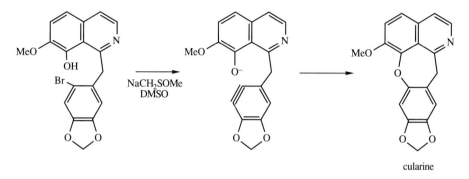

Synthesis of the alkaloid cularine

The second example, which again unfortunately proceeds in low yield, is concerned with a synthesis of lysergic acid. The key step in establishing the tetracyclic structure is the intramolecular addition of an enolate, formed by deprotonation of an α,β-unsaturated ester at the γ-position, to an aryne formed at the 4,5-position of a 3-substituted indoline.

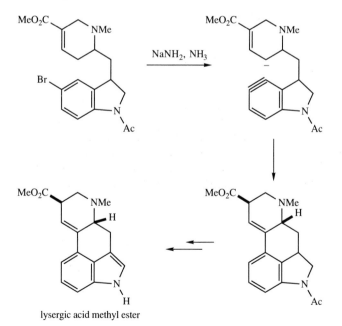

lysergic acid methyl ester

Intramolecular addition to an aryne as a key step in the synthesis of lysergic acid ester

Problems

1. Write a mechanism for the following transformation (* denotes the position of isotopic labelling):

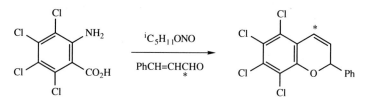

2. Predict the product of the following reaction:

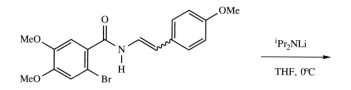

3. Write a mechanism for the following reaction of *ortho*-benzyne.

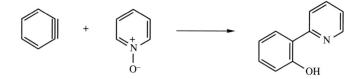

4. Suggest the likely product in the following reaction of *ortho*-benzyne:

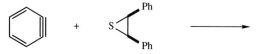

5. Write a mechanism for the following reaction:

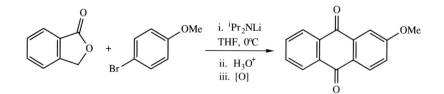

Further reading

Dehydrobenzene and cycloalkynes, R. W. Hoffmann, Academic Press, New York, 1967.

Frontier orbitals and organic chemical reactions, I. Fleming, John Wiley, Chichester, 1976.

Hetarynes, M. G. Reinecke, *Tetrahedron,* 1982, **38**, 427.

Reactive molecules: the neutral reactive intermediates in organic chemistry, C. Wentrup, Wiley, New York, 1984.

The stabilisation and reactivity of strained cyclic alkynes on transition-metal centres, M. A. Bennett, *Pure Appl. Chem.,* 1989, **61**, 1695.

Index